TOUR GUIDE TO OLD FORTS OF MONTANA, WYOMING NORTH & SOUTH DAKOTA

HERBERT M. HART

PRUETT PUBLISHING CO.
Boulder, CO

THE OLD ARMY PRESS
Ft. Collins, CO

Copyright © 1980 by Herbert M. Hart
All Rights Reserved

ISBN: 0-87108-570-4

Photo Credits:
 1. The National Archives
 5. State Historical Society of North Dakota
 6. Minnesota Historical Society
 8. Joslyn Art Museum, Omaha, NE
 9. Wyoming State Archives & Historical Dept.
30. South Dakota Historical Society
33. Montana Historical Society
Cover photo by Clark Babcock

P

PRUETT PUBLISHING CO.
3235 Prairie Ave.
Boulder, CO 80301

THE OLD ARMY PRESS
P.O. Box 2243
Ft. Collins, CO 80522

MONTANA

FORT CUSTER *(1)*

Fur trading brought the first forts to Montana and the banks of the Missouri and Yellowstone are dotted with so many that keeping them straight is an almost impossible task. Names and sites changed so frequently and were recorded by such a variety of travelers, traders and writers that this section must be viewed as a calculated guess in many cases.

Identifying the military posts is an easier matter, but this, too, is complicated by the many camps of the campaigns of the second half of the nineteenth century. This section presents most of the military and non-military forts of Montana with the request that local experts will come forth to correct any errors which have been committed in interpreting and evaluating the many conflicting sources.

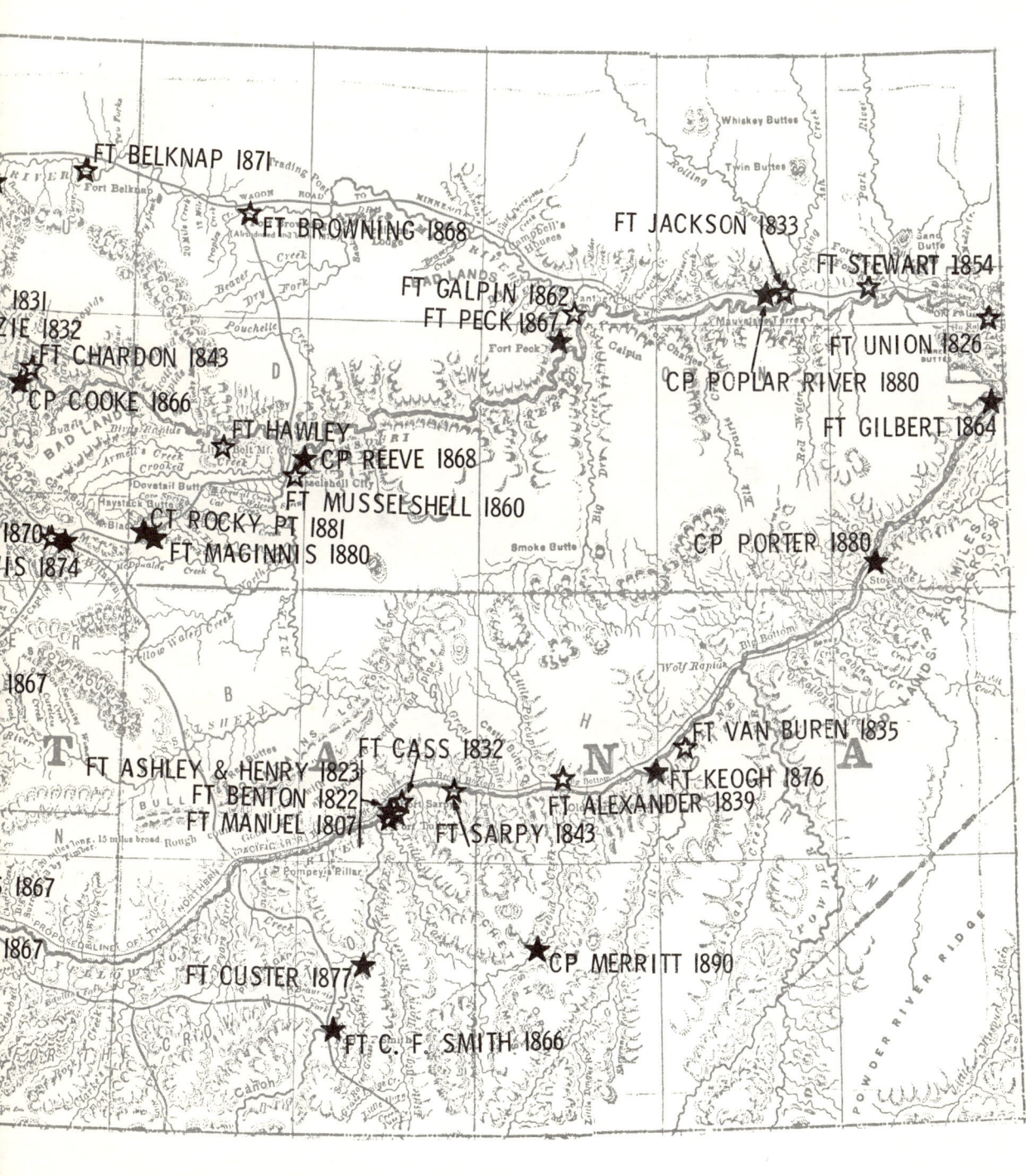

FORT BENTON *(1)*

Fort Alexander. 1839-50. A fur post built by Alexander McKenzie and Charles Larpenteur, this stockaded work was on the Yellowstone river opposite the mouth of the Rosebud. From Miles City, go west on I-94 for 33 miles. Turn right (north) and cross the Yellowstone river to the approximate site in the vicinity of Cartersville.

Fort Ashley and Henry. 1823-24. This trading post was near the site of Fort Manuel.

Fort Assinniboine. 1834-35. This 100-foot square temporary post was located on the Yellowstone river at the point where the steamboat *Assinniboine* went aground in the summer of 1834. Its site is unknown but its significance is that it marked the first advance of steamboats beyond the mouth of the Yellowstone.

Fort Assinniboine. 1879-1911. The largest military post constructed in Montana, this was a fort in the grand style. Long rows of brick buildings, castle-like towers at their corners, surrounded an immense parade ground. One of the major buildings and many other smaller ones are left at what is now a U.S. Agricultural Experiment station. From Havre, go south on US 87 for 7 miles to a gravel road. Turn left into station, 1 mile.

Camp Baker, *Fort Logan.*

Fort Belknap. 1871-86. This was a trading center that provided its name to the Indian agency 30 miles eastward. The site is in the vicinity of Chinook, on US 2 about 21 miles east of Havre.

Fort Benton. 1846-81. *Fort Lewis.* A fur post of the American Fur Company, Benton was taken over by the Army in 1869. The adobe trading post was inadequate for troop purposes. Several walls of the old post have been stabilized and an excellent museum tells the story of the post in the town park at the site. The town of Fort Benton is at the head of steamboat navigation of the Missouri river, 41 miles southeast of Great Falls off of US 87.

FORT ASSINNIBOINE *(1)*

FORT ASSINNIBOINE

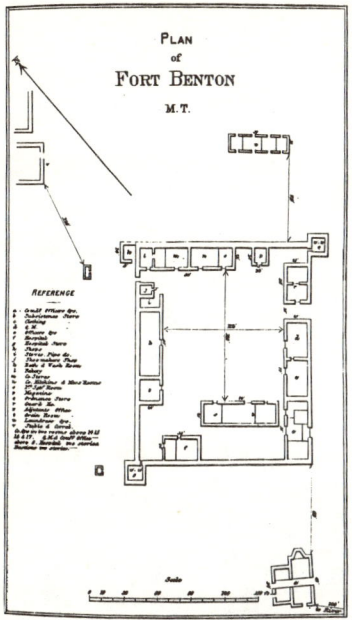

Fort Benton. 1822-23. This fur post was built by Joshua Pilcher at or near the earlier Fort Manuel.

Big Horn Post. *Fort Custer.*

Camp Bridger Pass. 1867. Montana militiamen put up this barricade to protect the Gallatin Valley from hostile Indians after the murder of John Bozeman in 1867. Two other temporary posts were Forts Howie and Elizabeth Meagher. The camp was at Bridger Pass which can be reached from Bozeman city on County route 293.

Fort Browning. 1868-72. This was a trading post. From Malta, go west on US 2 for 18 miles to Dodson. The site was 2 miles west of Dodson on the Milk river.

Fort Brule. *Fort McKenzie.*

Fort Campbell. 1845-60. Originally a wooden stockade, this trading post was first on the south bank of the Missouri near Fort Benton but in 1847 was moved across the stream and closer to Benton. It was built of adobe after the move, causing it to be the first adobe building in Montana. The approximate site is 1 mile west of Fort Benton on the river bank.

Fort Cass. 1832-35. *Tulloch's Fort.* Said to have been 130 feet square, of sapling cottonwood pickets with two bastions at extreme corners, this trading post was built by Samuel Tulloch. To reach the approximate site, go north from Bighorn, on I-94 about 44 miles west of Forsyth, for 2 miles to the south bank of the Yellowstone river.

Fort C.F. Smith. 1866-68. The Wagon Box Fight, one of the highlights of Western military history when 31 soldiers held off an estimated 800 Cheyennes, took place eight miles from this isolated Bozeman Trail fort. Under siege almost continually, this 300-foot square stockade was unable to communicate with the outside world for six months in 1866-67. From Hardin, take county 313 to Xavier, 23 miles. From here the route is a fair weather road to Yellowtail Dam. Within sight of the fort site and about a mile short of it, dirt road turns off to the rectangle of low mounds marked by a Boy Scout-placed stone memorial. The property is on Crow Indian tribe land.

Fort Chardon. 1843-45. Trader Alexander Culbertson built this short-lived post and then moved it closer to Fort Benton, changing the name to Fort Lewis. This site can be reached by following the directions to Camp Cooke but continuing north on county 236 across the free ferry. The Fort Chardon site was in the vicinity of the ferry landing on the north bank of the Missouri river.

Fort Connah. 1847-72. This was the last Hudson Bay post established in the United States.

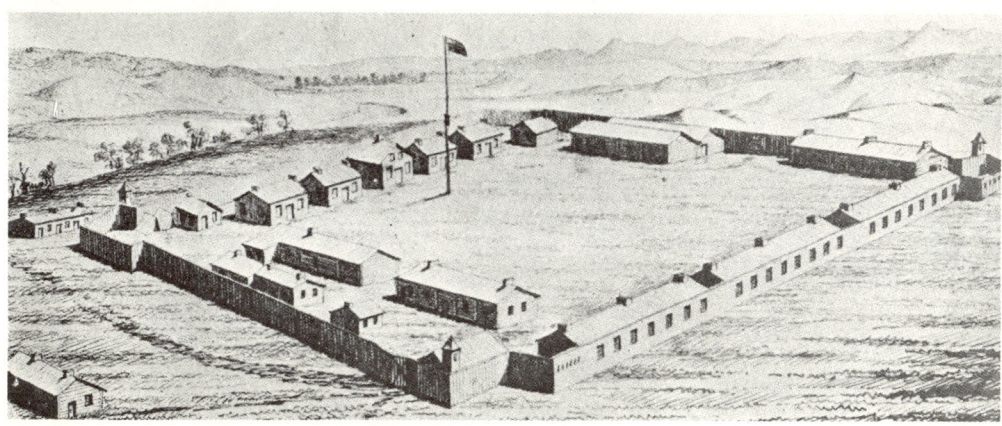

FORT C.F. SMITH *(1)*

The post remained an important trade center until 1872. State historical marker 45 stands on the site near an old stone house from trading days. From Missoula, go north on US 93 for 39 miles. At this point 6 miles north of St. Ignatius is the marker; the fort site is ¼ mile to the east (right side).

Camp Cooke. 1866-70. Indians were active against this stockaded post, scattering the horse herd at least once and later attacking the fort when the troops were away. Almost all traces have disappeared but there are evidences of excavations when the underbrush is not too heavy. From Lewistown, take Montana 19 north for 15 miles to Hilger. Turn left on gravel road, go 23 miles to Lohse Ferry, free ferry across Missouri. Half mile short of ferry landing is a building remaining from Power-Norris store, dating from last half of 19th century. Behind this building is the mouth of the Judith river. Cooke site can be reached by wading the Judith at this point and walking westward 1 mile.

Camp Cummings. 1867. This was a militia protection against the Indian scare in conjunction with outposts at Bozeman and Bridger Passes. This site was in Virginia City, then the capital of Montana Territory but now a lively ghost town.

Fort Custer. 1877-98. *Big Horn Post.* Scattered cellars surrounding a DAR marker indicate where more than 1,000 soldiers were stationed to keep peace after and 15 miles away from the Custer Massacre of 1876. Most of the buildings served as the nucleus of nearby Hardin after the elaborate post was dismantled. From Hardin, go south on US 87 for 2 miles. After crossing Bighorn river, turn right (west) on dirt road that climbs to top of bluff and site of fort where the marker, cellars, and a golf clubhouse are the only evidences.

Fort Elizabeth Meagher. 1867. Named after the wife of the former acting governor of Montana Territory, this stockade was built by the militia during the Indian scares following the murder of John Bozeman. Take I-90 for 6 miles east from Bozeman to the approximate site near the mouth of Rock Creek.

Fort Ellis. 1867-86. Soldiers were dispatched from this key military fort throughout the post-Civil War period; its troops were part of the 1876 three pronged-pincer against the Sioux and Cheyenne that was blunted by the Custer Massacre. From Bozeman, go east on I-90 for 3.4 miles to marker on left (north) side of highway. The Fort Ellis Experiment Station now occupies the site.

Fort Fizzle. 1877. This series of entrenchments and barricades were erected by men of the 7th U.S. Infantry and local volunteers when word reached Missoula that the Nez Perce Indians were fleeing in their direction. The Indians so outnumbered the defenders when they arrived on July 28, 1877, that the temporary fort's defenses fizzled— hence the uncomplimentary name. From Lolo (on US 93 about 13 miles south of Missoula), take US 12 for 5 miles west (right) of

CAMP COOK *(33)*

FORT CUSTER *(1)*

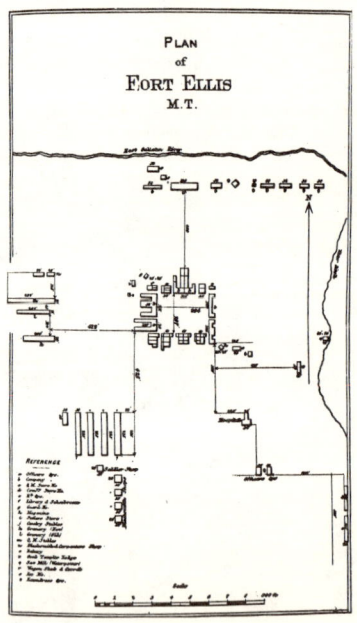

8

FORT KEOGH (1)

Lolo to where the Lolo Trail enters the Bitterroot Valley.

Fort Galpin. 1862. This was a trading post near the mouth of the Milk river on the Missouri. The site is in the vicinity of modern Fort Peck.

Fort Gilbert. 1864-67. This was a trading post at the edge of the Fort Buford, N.D., reservation, suggesting that one of its economic interests may have been the extracurricular attractions unavailable on the military reservation. From Glendive, go north on Montana 16 for 52 miles to Sidney; 5 miles north of Sidney on Montana 200 is State Historical marker 104 indicating that the fort site is east of the road marker on the bank of the Yellowstone river.

Fort Harrison. *Fort William Henry Harrison.*

Fort Hawley. This fur fort site can be reached only via a considerable overland trek. Follow the directions to Fort Musselshell. From this site, Fort Hawley location can be reached by swimming across Musselshell river and following river to its mouth and intersection with the Missouri river. Follow the bank of the Missouri around to north and west to approximate fort site, approximately 25 miles directly east of US 191 crossing of the river.

Helena Barracks. 1877-78. This was the campsite of troops sent to maintain the peace in Helena. They were quartered at the Helena fairgrounds.

Fort Howie. 1867. This was another militia stockade erected for the Indian scare of 1867. The general site is on the bank of the Yellowstone near the mouth of Shields river. This would put this temporary post approximately 7 miles east of Bozeman.

Fort Jackson. 1833-34. This 50-foot square post was built by trader C.S. Chardon as a wintering-over camp. The approximate site is on the Missouri river at the mouth of the Poplar in the vicinity of the town of Poplar.

Cantonment Jordan. 1859-60. This was a winter camp for the construction crew of military road builder Captain John Mullan while he was pushing a wagon road into the Pacific Northwest. From DeBorgia (83 miles west of Missoula),

FORT KEOGH, Company Quarters (1)

go 2 miles east on I-90 to State Historical Marker 11 at the site.

Fort Keogh. 1876-1908. *Cantonment on the Tongue River, New Post on the Yellowstone, Tongue River Barracks.* The first site of Fort Keogh was known as Tongue River Barracks and it was from this site next to the Tongue river that General Nelson A. Miles fielded troops in 1876. Later the elaborate new post was built a mile to the west.

Only a well marks the first site; two old officers quarters are at the second site, used by the U.S. Range Livestock Experimental Station.

From Miles City, take Main street west to the edge of town. On the right is the Range Riders Museum and, next to it, one of the officers quarters moved from Fort Keogh. Beyond, in the field next to the river, is the site of Tongue River Cantonment.

Further west 1 mile is the Livestock Experimental Station, site of Fort Keogh. Until destroyed by fire in 1973 the quarters occupied by General Miles were in use at the apex of officers' row.

Fort Kipp. *Fort Stewart.*

Fort LaBarge. 1862-63. The St. Louis firm of LaBarge, Harkness and Company intended this inclosure—300 feet long by 200 wide—to be their principal trading post and named it after one of their founders, the leading Missouri river steamboat captain. After a year of operation, the post was sold to nearby Fort Benton. Follow directions to Fort Benton; Fort LaBarge was west of Fort Benton in the town.

Camp Lewis. 1874. This short-term temporary post was once to the west of Lewistown, but now has been swallowed up by the city. The site is on the western edge of the city.

Fort Lewis. *Fort Benton.*

Fort Lewis. 1845-46. This trading post was on the right bank of the Missouri river opposite Pablois Island, 18 miles above the Fort Benton bridge. This fort was torn down in 1846 and moved to the site of Fort Benton. From Fort Benton go south on US 87 about 14 miles to Carter and the intersection with an improved road. Turn south (left) and cross river. The approximate site was on the south bank of the river.

Fort Lewis. 1846-50. Fort Benton's first name until this trading post was torn down in 1850, Fort Lewis was then built of adobe and rechristened Fort Benton on Christmas Day, 1850, Follow directions to Fort Benton.

Fort Lisa. *Fort Manuel.*

Fort McKenzie. 1832-43. *Fort Brule.* This 140-foot square trading post fell victim to alcohol rather than Indians, although both were involved. It was the reduction in the whiskey ration that increased Indian restlessness and forced abandonment of this prominent fort. Afterward, the Indians burned the deserted stockade; in later years, the remnants often were called Fort Brule. From Fort Benton, go north on US 87 for 11 miles to Loma. The approximate site of Fort McKenzie is on the north bank of the Missouri about 6 miles south of Loma.

Camp Merritt. 1890-98. Providing protection to the Tongue River Agency in the latter stages of Indian difficulties was the mission of this post on Lame Deer Creek in what is now the Northern Cheyenne Indian reservation. From Hardin, take I-90 south 15 miles to US 212 (not far from the Little Big Horn battlefield). Turn left (east) for 42 miles to Lame Deer, vicinity of the site of Camp Merritt.

Fort Missoula. 1877-1946. Troops from Missoula participated in the nearby Battle of the Big Hole in 1877 and in a second battle with the Nez Perce in 1878. From Missoula, take US 93 southwest to the edge of the city. Road to the fort branches off to the right at the city limits. Some period buildings are left amidst modern construc-

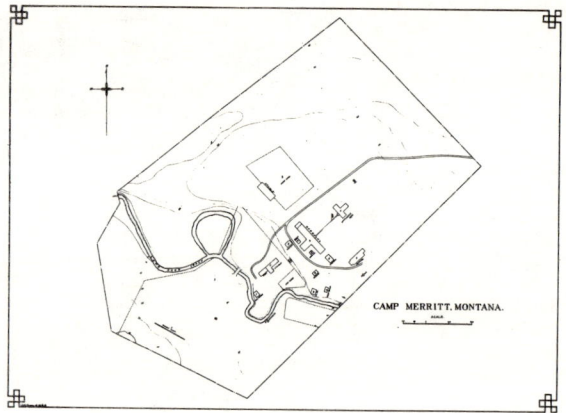

Plan of Camp Merritt *(1)*

FORT MISSOULA *(1)*

FORT LOGAN (33)

tion used by reserve units and government agencies. The former quartermaster building is a State Historical Society museum and a non-commissioned officer's quarters has been renovated by the Western Ghost Town Preservation Society.

Fort Logan. 1869-80. *Camp Baker.* This horseshoe-shaped post included a blockhouse that is still at the site. Known as Camp Baker for most of its active life at two sites, Fort Logan's second location has many buildings in modern use as farm residences or utility sheds. From White Sulphur Springs, take asphalt road past post office west of town. In about 1 mile, turn right on gravel road. Go 19 miles to Byrd Ranch, privately owned site of the fort from 1870 to 1880. First location, 1869-70, was ten miles north on the west bank of the Smith river. There are no traces here. Baker name was given to first temporary post, honoring commander, Brevet Colonel Eugene M. Baker; it was renamed to memorialize Captain William Logan, killed in the Battle of the Big Hole, in 1878.

Fort Maginnis. 1880-90. Both Indian and civilian rustlers were the main targets of this large post that was built after the real Indian wars were over. All of the buildings have been dismantled but the foundations still are prominent and, weather depending, marked. From Lewistown, take US 87 east 14 miles to gravel road leading to ghost town of Giltedge. About 1 mile short of Giltedge (buildings can be seen in distance), turn right and follow winding road about 9 miles to site of fort. Small markers appear at road junctions, but do not depend on this. Fort is across Forge Creek, left side of road. Do not attempt to reach site from the north; despite oil company map indications, this route is closed.

Fort Manuel. 1807-11. *Fort Lisa.* The first building in Montana, this fur post was erected on the site of Lewis and Clark's camp of July 26, 1806; seventy years later, the same site was the jumping off point for General Terry's column's unsuccessful attempt to contact Custer before the massacre. State marker 22 is 1 mile west of Bighorn off of I-95 on the south side of the Yellowstone river.

Camp Morris. 1883. This temporary post was officially listed as being on the "west side of Cottonwood Creek near the Sunset Grass Hills." From Shelby, go north on I-15 for 15 miles to Oilmont exit. Turn right (east) and follow gravel and dirt road for about 43 miles of miscellaneous turns to Whitlash, the approximate site of Camp Morris.

Fort Musselshell. 1860-70. This trading post catered to Indians and river workers during its active days; its vicinity was a hangout for rustlers until Vigilantes took over. From Lewistown, go east on US 87 for 30 miles to the point where Montana 200 branches off. Continue on Montana 200 for 55 miles to Mosby. At Mosby, take unimproved road north along the Musselshell river about 35 miles to the Missouri river and the approximate site of the post. State marker 66, 1.5 miles east of Mosby, tells about this fort.

Fort Owen. 1850-59. Father DeSmet built St. Mary's Mission at this site in 1841 but Indian troubles caused his missionaries to sell it to Major John Owen in 1850 for use as a trading post. One building is left. From Missoula go south on US 93 for 28.5 miles to a gravel road. Turn left (east) on gravel road across the Bitterroot river bridge .5 mile.

Fort Peck. 1867. This trading post was a 300-foot square stockade 12-feet high that boasted three bastions on the front and two along the rear. Both soldiers and Indian peace commissioners used it as a base camp. Montana marker 105 tells the story 1.5 miles west of the modern town of Fort Peck on Montana 24. The actual site of the fort has been inundated by Fort Peck reservoir; it was about 1 mile west of the Fort Peck dam.

Fort Piegan. 1831-32. James Kipp built this trading post for the winter. As soon as he left it, after a successful beaver trading season, the fort was burned by the Indians. From Fort Benton, go north on US 87 to Loma, 12 miles. Montana marker 43, ¾ miles south of Loma, tells the story of this fort and also of the Lewis and Clark camp of June 3, 1805. A dirt road, to the right (east), leads to the intersection of the Missouri and Marias river, site of Fort Piegan.

Camp Poplar River. 1880-93. This tiny post has disappeared except for the fact that the town of Poplar, on the site, bears the same name. The post was 2 miles north of the Missouri on the south bank of the Poplar river.

Camp Porter. 1880-81. This single-year camp provided protection to Northern Pacific railroad construction crews. The town of Glendive now occupies the site on I-94.

Fort Raymond. *Post Three Forks.*

Fort Reed. 1870. The decade of the 1870's saw this active trading post in operation amidst favorite Indian hunting grounds. Montana marker 93 tells the story on US 87 about 1 mile west of Lewistown.

Camp Reeve. 1868. This stockaded post was garrisoned by troops from Camp Cooke during the summer of 1868, providing protection for the settlers who tried to establish the town of Musselshell. The project was sponsored by the Montana Hide and Fur Company, had eight buildings and 50 inhabitants, but was abandoned when the Indians rustled the stock and killed two soldiers. The post consisted of tents surrounded by a stockade south of the town site. Follow directions to Fort Musselshell, the approximate site of the town.

Camp Reynolds. *Fort Shaw.*

Cantonment Rocky Point. 1881. This was the construction camp for nearby Fort Maginnis and was abandoned when the fort was completed.

Fort Sarpy. 1843-60. Fifteen-foot high pickets were set into a 100-foot square stockade to protect this final Crow Indian trading post of the American Fur Company. From Forsyth, go west on I-94 for 26 miles to Hysham turnoff. Go north to Yellowstone river; Fort Sarpy was on the south bank of the river at this approximate location.

CANTONMENT STEVENS, 1859
From Railroad Survey Report *(8)*

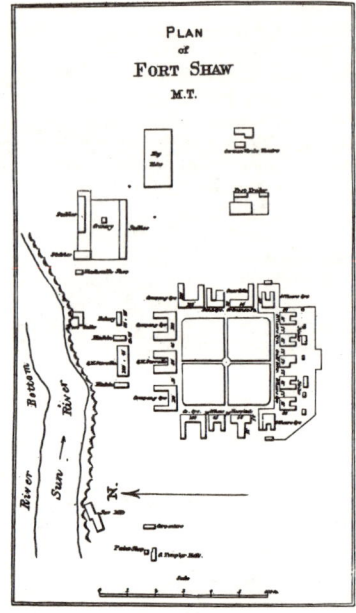

Fort Shaw. 1867-91. *Camp Reynolds.* General John Gibbons took his garrison to the Little Big Horn in 1876 but was too late to prevent the Custer massacre. Built around a 400-foot square, many of the adobe and shingled buildings continue to be used by the county school system. From Great Falls, go west on I-15 and US 89 about 19 miles to the intersection with Montana 200. Follow this for 5 miles to the town of Fort Shaw. The cluster of fort buildings is a half mile north of the town next to the Syn river.

Fort Sherman. 1873-74. This was a trading post at the site of what later became Lewistown. The buildings were dismantled when trade proved unprofitable.

Cantonment Stevens. 1853. This post was so-called after Governor Isaac I. Stevens of Washington, who stayed here in 1859 with Major Owen of Fort Owen and Lieutenant John Mullan, the road builder, during their railroad survey expedition. It included four log buildings, a corral and tents. Mullan's initiative in establishing and erecting the camp drew praise from Stevens who noted that now the government would save money by not having to rent quarters. The site is the location of the community of Stevensville, about 1 mile south of the turnoff to Fort Owens.

Fort Stewart. 1854-63. *Fort Kipp.* Actually two trading posts were established at the same site, the Kipp post coming first and followed by Stewart about 200 yards away. By 1860 both had been burned down and Stewart was rebuilt on the Kipp site because usable chimneys were still standing. From Poplar, go east on US 2 for about 23 miles; walk south across railroad tracks ¾ mile

12

to open plain 1 mile north of Missouri river (where the river makes so-called Devil's Elbow).

Post Three Forks. 1810. *Fort Raymond.* This Lewis and Clark campsite was used by the St. Louis Missouri Fur Company as the location of a 300-foot square double stockade of logs but Indian trouble forced the abandonment of the post after a single summer of operation. From Bozeman, go west on I-90 for 28 miles to Three Forks turnoff. Montana marker 14 gives the history of the Three Forks of the Missouri, 1 mile east of the town of Three Forks. The fort site was on the bank of the Missouri, north of the town, and is believed to have been washed away.

Cantonment on the Tongue River. Tongue River Barracks. *Fort Keogh.*

Tulloch's Fort. *Fort Cass.*

Fort Union. 1826. This trading post was the first Fort Union and is one reason that the famous North Dakota fort is often erroneously said to be in Montana. From Glasgow, go east on US 2 for 28 miles to Frazer, the approximate site of the fort on the banks of the Missouri.

Fort Van Buren. 1835-43. The second American Fur Company trading post on the Yellowstone, Fort Van Buren was burned after its abandonment. The approximate site is probably the same as the first Cantonment on the Tongue River, the early version of Fort Keogh, next to Miles City on I-94.

Fort William Henry Harrison. 1895-1913. The Veterans Administration now occupies this last of many Montana forts, built originally as part of the scheme to consolidate troops into larger camps from which they could be rushed by railroad to the trouble spots. The fort is in Helena.

Cantonment Wright. 1861-62. This was the winter camp for Captain John Mullan's road builders. From Missoula, go east on I-90 for 17 miles to Milltown. The site is a quarter mile west of Milltown on the edge of the highway, marked by Montana marker 9.

New Post on the Yellowstone. *Fort Keogh.*

WYOMING

FORT BRIDGER *(1)*

Cattlemen provided Wyoming with a war that was a little different from that of the other Western states and the camps of the Johnson County War are included in this listing, in addition to those of the military, traders and settlers throughout the nineteenth century.

The geography of Wyoming called for posts along the trails and in the mining camps that could be independent when weather or warfare dictated. This geography also meant an isolation that has preserved the sites of many of the posts and many of the remnants, but it is not uncommon to come upon a lone marker standing amidst sage brush many miles from civilization, vague surface indentations nearby serving as the sole evidence of early occupancy.

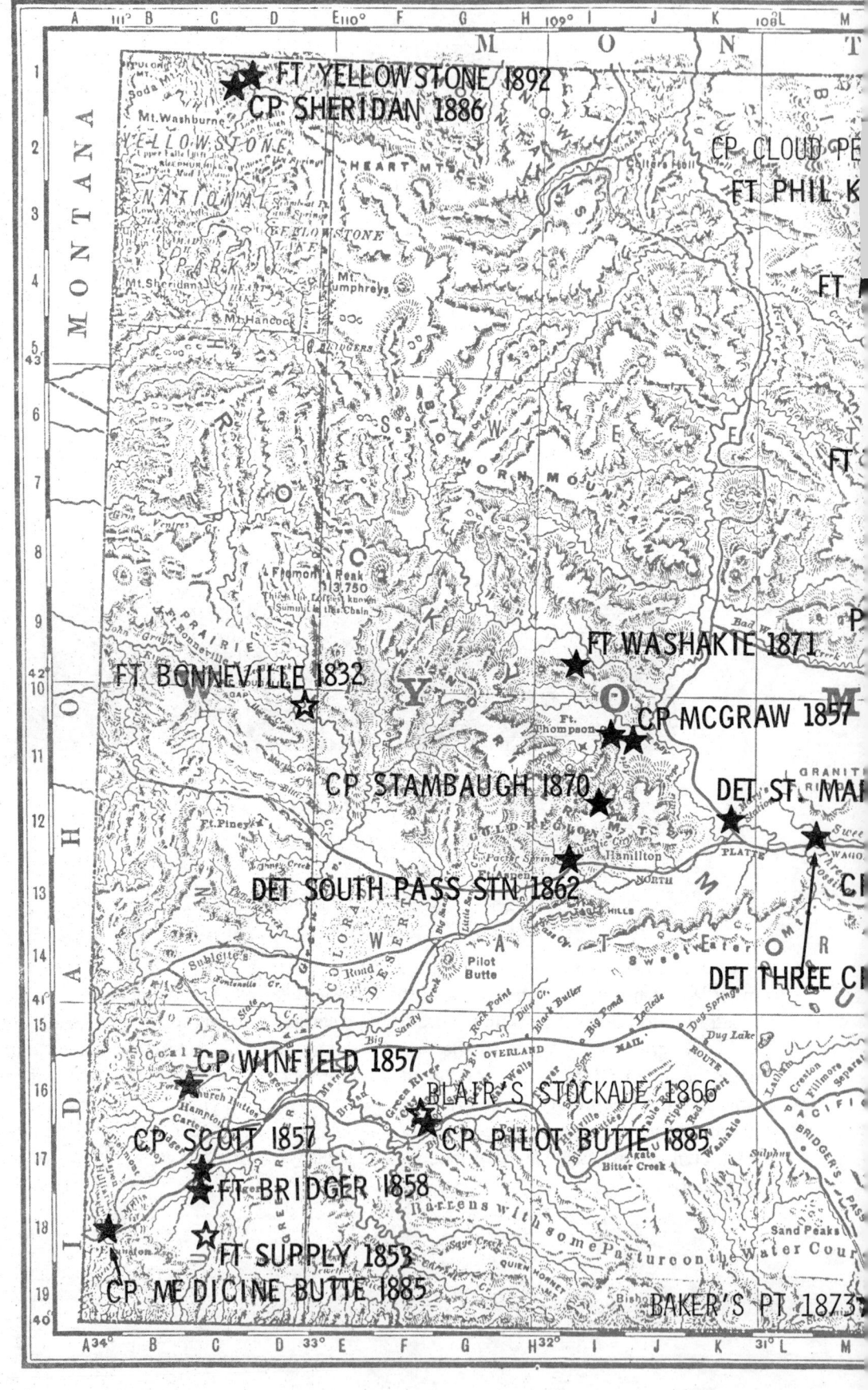

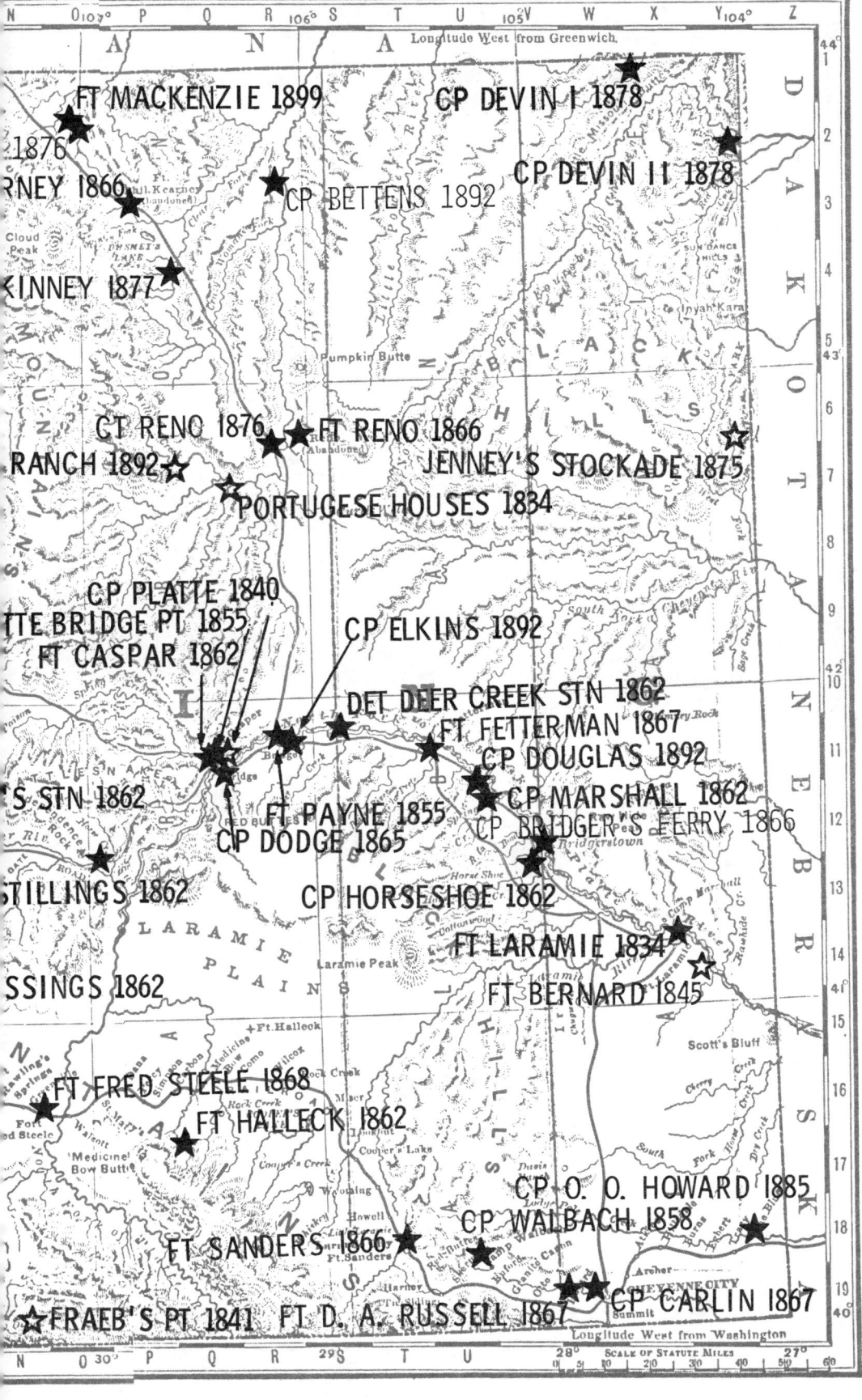

Camp Augur. 1869-71. *Camp Brown.* Log cabins with sod roofs were surrounded by a ditch within an enclosure 175 feet by 125 at Camp Augur. The post was intended to protect both Indians and settlers in the Sweetwater mining district. It was renamed Camp Brown in 1870 and moved to Little Wind river where it would ultimately be known as Fort Washakie. The site is within the city of Lander (at the US routes 26 and 287) on Main street near 3d street. A block of granite marks the spot.

Baker's Post. 1873. Not a trading post, this three-story blockhouse was the retirement home of Mountain Man James Baker, one of the survivors of the battle at Fraeb's Post, 10 miles to the east in 1841. The building was moved to Frontier Park in Cheyenne in 1917, about 30 years after Baker's death. The site is in a field 1.2 miles south of Savery (see Fraeb's Post item for detailed directions).

Fort Bernard. 1845-46. *Richards Post.* Whiskey and price-cutting were the tools used by John Baptiste Richard to compete with nearby Fort Laramie. This trading post was the successor to Fort Platte and, because it had inherited Platte's inventory, its trade was brisk even before construction was finished on the rough log and brick building. Bernard's accommodations were termed "far inferior to those of an ordinary stable" but this was not the reason for its short life. Apparently Richards' methods and illicit whiskey trade were sufficient reason for the post to be burned, whether by competition or drunken patrons has never been settled. From Fort Laramie town, go southeast on US 26 for 7.3 miles; across the river from here (to the right) is the site of Fort Bernard.

Camp Bettens. 1892-95. Six companies of the 6th Cavalry were deployed here temporarily from Fort Robinson in the later days of the so-called Johnson County War. They were at the camp from mid-June to mid-November, 1892, and returned for portions of the next three summers when the site served as a Camp of Instruction. The site is near Arvada. From Sheridan, go southeast on US 14 for 59 miles to a right (south) turn to Arvada, 3 miles.

Blair's Stockade. 1866-70. This was a trading post of Archie and Duncan Blair, now virtually impossible to locate because of time and construction. From Rock Springs, go north on US 187 for .7 mile to a dirt road. Turn left .2 mile to the site on the west side of Killpecker creek at the foot of the chalk bluffs.

Fort Bonneville. 1832. *Fort Nonsense.* Captain Benjamin L.E. Bonneville spent several weeks and considerable effort in building this log enclosure, blockhouses at two diagonally opposite corners. The post was hardly finished before he moved to a succession of temporary camps in this area of fur trade rendezvous sites. The wasted work won for the abandoned fort the second name listed. From Rock Springs go north on US 187 for 110 miles to the intersection with US 189 north of Daniel. Head straight west 4 miles to point where marker stone and information sign tell the story of Fort Bonneville at the left (south) side of the road.

Fort Bridger. 1842-57. This was the trading post of Jim Bridger until he left in 1853 and, beginning in 1854, of Mormon traders representing Brigham Young. Bridger's original post was the usual collection of log huts connected by mud daubed pickets. In 1855 the Mormons erected a 400 foot square cobblestone wall and added some log buildings within the fort, all of which they destroyed when the U.S. Army and the Utah Expedition arrived in 1857. Remnants of the Mormon cobblestone wall are part of the Fort Bridger State Monument next to the town of Fort Bridger on I-80 about 32 miles east of Evanston.

Fort Bridger. 1858-90. After a cold winter in nearby Camp Scott, the Army moved on to Utah but left a detachment to rebuild a new Fort Bridger, by 1858 designated an official Army post. Log buildings were erected around a parade ground and gradually the post became an expansive cluster of military buildings for traveler protection. Fort Bridger State Monument has restored some buildings, maintained or marked the others, and provided a museum that tells the

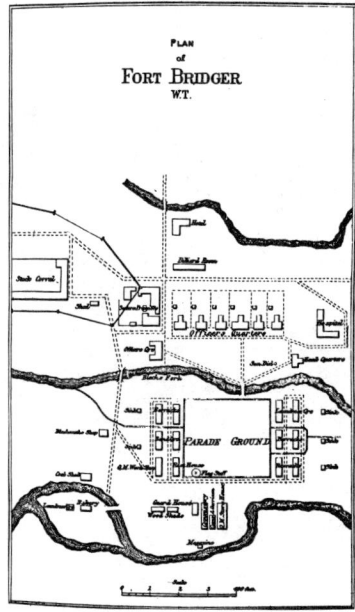

story of the three phases of the post's life: trading post, Mormon fort, and Army garrison.

Detachment at Bridger's Ferry. 1866. Galvanized Yankees of Company I, 11th Ohio Cavalry, spent the summer of 1866 at this field camp guarding Benjamin Mills' ferry across the North Platte river. Howitzers were placed at each landing. With 21-man detachment, the post stopped the raids on the ferry while it was used to supply and reinforce the expeditions and new posts to the north.

From Douglas, go southeast on I-25 to a point 1.5 miles south of Orin. There is a marker indicating that the ferry site is on the river banks north of the railroad tracks.

Camp Brown. *Camp Augur, Fort Washakie.*

Camp Carlin. 1867-88. *Cheyenne Depot, Quartermaster Depot near Cheyenne.* Adjacent to the Fort D.A. Russell reservation, Camp Carlin was the supply center for the posts to the north and west. It had wooden quarters for one company, a guardhouse, three officers quarters, and 16 portable store houses. From Cheyenne, take Randall avenue west 1 mile from the capitol. Obtain pass to Warren Air Force Base at the gate, then turn left at 1st street. A granite marker near the railroad marks the site.

and also on his way back after the battle. The site is now within the city of Sheridan where it is marked in the city park.

Fort Connor. *Fort Reno.*

Crook Supply Camp. *Camp Cloud Peak.*

Camp Davis. *Fort Paine.*

Fort D.A. Russell. 1867-1948. *Fort Francis E. Warren, Post of Crow Creek.* An unusual diamond shaped parade ground flanked by 11 wooden barracks and 15 officers quarters was a feature of this headquarters fort. Measuring 1,040 by 800 feet, the parade ground and many of the buildings, rebuilt with brick in 1880, are still in use at Warren Air Force Base, 1 mile west of Cheyenne. The post cemetery has graves removed from many abandoned forts.

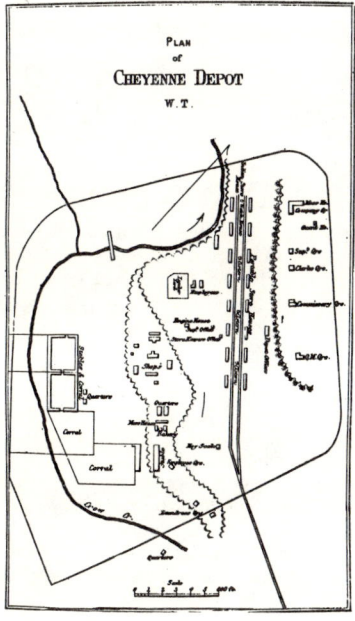

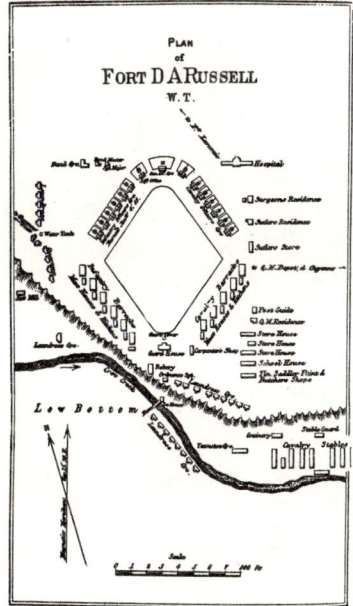

Fort Caspar. 1862-67. *Platte Bridge Station.* Troops were assigned here to guard the 1,000-foot bridge across the Platte during and after the Civil War. This was at a crossing 5 miles above John Richard's bridge (see Camp at Platte Bridge). After Lieutenant Caspar Collins was killed in a battle with Indians, the post was renamed in his honor. Portions of the log fort have been reconstructed on the original foundations and are maintained by the city of Casper (named, but with altered spelling, after the fort). From Casper, take 13th street west 1.5 miles to the fort. Pilings from the bridge are north of the fort.

Camp Clay. *Fort Paine.*

Camp Cloud Peak. 1876. *Crook Supply Camp.* General Crook camped here from June 10 to 16, 1876, on his way to the Battle at the Rosebud

FORT CASPAR (Platte Bridge) *(1)*

FORT D.A. RUSSELL

Detachment at Deer Creek Station. 1862-66. Details of troopers watched over the Oregon Trail travel at this station and, when the telegraph was in operation, protected both the construction crews and line tenders. From Casper, take I-25 for 18 miles east to Glenrock. At Glenrock, take a gravel road from the Deer Creek bridge .3 mile to Glenrock Park. The site of the station is in this park.

Camp Devin. 1878. Building the telegraph line to Fort Keogh meant two temporary field sites for this summer post. The first post was established on June 30, 1878, on the Little Missouri river. On August 10 the post was moved to Oak Creek south of the Belle Fourche river where the five companies of the 3d U.S. cavalry were headquartered until September 22. The locations of both sites are only approximate.

To visit both camps, go north on I-90 for 13 miles to the intersection with road number 0603 to Alladin, 9 miles. Camp Devin I can be reached by turning left (west) on Wyoming 24 at this point for 25 miles to Hulett. At Hulett, inquire locally for dirt road that goes north to the Little Missouri and the site near the Montana border.

Camp Devin II can be reached from Alladin by inquiring locally for the road to the junction of Oak Creek on the Belle Fourche, about 10 miles north of Alladin.

Camp Dodge. 1865. The spring attacks by the Indians on most of the Oregon Trail stage and telegraph stations caused the 11th Kansas Cavalry to be deployed along the line with headquarters at this temporary post southeast of Platte Station on the east side of Upper Garden Creek. Between April 19 and June 28, 1865, several patrols were fielded from here until the headquarters was shifted to Camp Marshall. Follow directions to Fort Caspar; four miles southeast of Caspar is the site of Camp Dodge.

Camp at Douglas. 1892. Troops A and K, 6th Cavalry, were stationed near Douglas for a short time in the fall, 1892, as an aftermath of the Johnson County War. Douglas is north of I-26 and 48 miles east of Casper.

Camp Elkins. 1892. Six companies of the 6th Cavalry, permanently stationed at Fort Niobrara, Nebraska, established this temporary field camp on June 20, 1892, during the final days of the Johnson County War. These troops were part of the overall plan by the Wyoming cattle barons against the small independent ranchers. The barons had attempted to obtain a declaration of martial law, something the White House refused to do, so President Harrison appeased his political friends by moving troops into the area. The site is between Fort Fetterman and Casper. Inquire locally at Casper for directions.

Fort Fetterman. 1867-82. The four barracks and seven officers quarters of this key fort were originally surrounded by a high plank fence but this gradually disappeared.

Were it not for the isolated position at the jumping off point for the Bozeman Trail, Fetterman would have been highly prized for assignment — at least for the officers who found the quarters particularly good. Apparently the post commanders kept building themselves new quarters, the rejected ones being turned over to the less senior officers. General Crook had his camp north of the fort prior to his 1876 expedition. From Douglas, take I-25 west one exit to road 0502, about 2.7 miles. Go north on 0502 for 6.8 miles to the fort site, marked on the right side (east) of the road where several buildings are left in a state park.

Fort Fred Steele. 1868-86. The unusual arc-shape layout of this railroad post is still obvious in the building foundations of the deserted hamlet of Fred Steele, once the garrison's quarters. The last building burned on New Year's Eve 1976. The site is a state park and preservation is planned. From Rawlins, go east on I-80 for 14 miles to a left (north) turn onto a gravel road just before reaching the North Platte river. The fort site is at the end of this road, about 2.5 miles, on the other side of a road under the railroad.

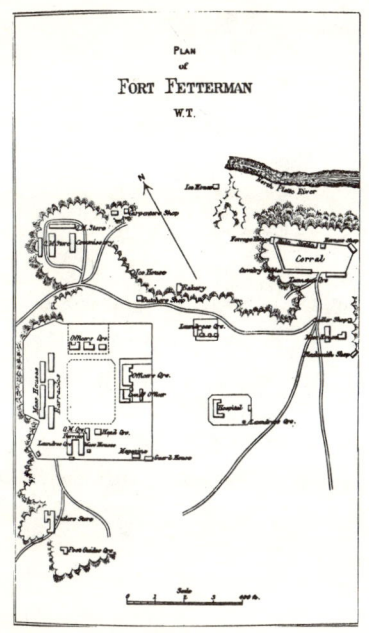
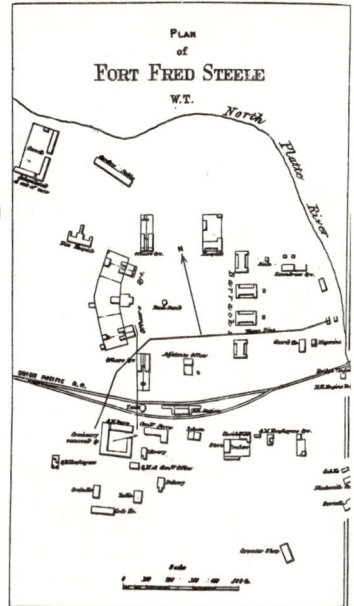

FORT FRED STEELE *(1)*

Fraeb's Post. 1841. Mountain man Henry Fraeb was one of five men killed at this site by Sioux and Cheyennes who resented his attempts to build a trading post here. The fighting took place amidst the logs felled to build the fort. The defenders were able to hold off the several hundred warriors even though Fraeb, mortally wounded, did much of his fighting sitting against a stump until he died. The site is within Battle Mountain State Game Preserve. From Rawlins, go west on I-80 for 26 miles to Wyoming 789. Go south on 789 the 51 miles to Baggs. Turn left (east) at Baggs through Dixon, 7 miles and the site of *Baker's Post,* to Savery, 4 miles. At Savery inquire locally for the unimproved road to the site, 4 miles to the east and just north of the Colorado border.

Fort Halleck. 1862-66. Elk Mountain looks down on the site of this protector for the railroad where several original buildings and the post cemetery are left. From Rawlins, go east on I-80/US 30 for about 40 miles to the turnoff to Hanna. Turn right (south) 10.5 miles to a dirt road. Turn right (west) 4 miles to another dirt road. Turn right (north) 1.7 miles to the entry road to the Quealy Ranch which presently owns and occupies the site and buildings of Fort Halleck.

Camp on Ham's Fork. *Camp Winfield.*

Camp Hat Creek. 1876-77. *Camp on Sage Creek.* This was a stockaded sub-post of Fort Laramie for protection of the Black Hills stage route. It was at the site of Camp on Sage Creek occupied in 1875 as part of the perimeter guard-

21

FORT HALLECK, from a drawing by bugler C. Moellman, CO. G, 11th Ohio Cavalry in 1863 *(1)*

ing the Black Hills from civilian miners. The site is next to the Hat Creek stage station on US 85 about 14 miles north of Lusk.

Camp Horseshoe. 1862-66. *Detachment at Horseshoe Station.* A lieutenant and 38 men were assigned to guard travelers and telegraph at this important station, the first stopping place for most stages from Fort Laramie. The stockade became a ranch in 1868. When the occupants fired on a passing group of Indians led by Crazy Horse, killing two, the Indians attacked and burned the place. The ranchers were spared because they hid in a root cellar. To reach the site go southeast on I-25 from Douglas to Glendo, 27 miles. Continue south 2.5 miles. At this point, the station site is on the left (east) where there is an historical marker.

Jenney's Stockade. 1875. An expedition into the Black Hills with neither military nor settler objectives was the geological survey led by government geologist Walter P. Jenney in 1875.

Correctly viewing the survey as the first step in a government plan to buy the Black Hills from the Sioux and open them to prospecting, miners were pleased to help Jenney put up this stockade to protect his scientists.

The site of the stockade is on a ranch near Newcastle in northeastern Wyoming. From Newcastle, go southeast on US 16 for 5.2 miles to the Stockade Beaver creek, site of Jenney's Stockade where there is an historical marker.

The stockade was moved in 1933 to Newcastle where it was rebuilt as a pioneer museum in the courthouse yard on Warren avenue.

Fort John. *Fort Laramie.*

Fort John Buford. *Fort Sanders.*

Detachment at La Bonte Station. *Camp Marshall.*

Fort Laramie. 1834-49. *Fort John, Fort John on the Laramie, Fort William.* This was one of the leading trading posts in the northwest, founded in 1834 as Fort William by the firm of William Sublette, Robert Campbell and others. It was a palisade 18 feet high, bastions on two diagonally opposite corners, with a few adobe buildings inside. After it was sold to Milton Sublette and James Bridger, the post was rebuilt of adobe and

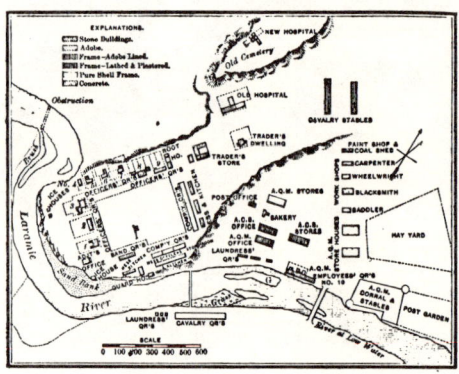

renamed Fort John. Despite all of this, the names of Fort Laramie or Fort John on the Laramie were used frequently because of their geographical description made it obvious what fort was concerned. The buildings were sold in 1849 and became the military post of Fort Laramie.

Fort Laramie. 1849-90. The Grattan Massacre that precipitated the Indian Wars of the 1850's took place near this critical military fort of the plains. From the stockaded trading post purchased in 1849, this fort became a large, rambling affair of more than 150 buildings. It is part of the National Park Service and the buildings have been restored, rebuilt, stabilized or site marked for visitor viewing. From Cheyenne, go north on I-25 for 82 miles to US 26. Turn right (east) for 27 miles to the town of Fort Laramie. A marked gravel road goes 2 miles to the Fort Laramie National Monument.

Fort Mackenzie. 1899-1918. Sheridan's military tradition amidst the forts of the Bozeman Trail and subsequent events became permanent with the establishing of this major fort at the end of the 19th century. The first troops arrived in 1899 but it was not until 1905 that the post was fully garrisoned. Although given up by the Army during the final week of World War I, the fort continues to serve the armed forces as a Veterans Administration hospital specializing in psychiatric care. The former fort is 2 miles northwest of Sheridan off of I-90.

Camp Marshall. 1862-66. *Detachment at La Bonte Station.* Under the La Bonte name, this post was guarded by detachments of troops until 1864 when Captain Levi G. Marshall, 11th Ohio Cavalry, supervised the construction of a more permanent fort. This was little better, being nothing more than a crude square of stables, barracks, and a stockade surrounding the telegraph station. A pair of mountain howitzers provided artillery defense. The site can be reached from Douglas by taking the road across the Platte to Esterbrook. At about 10 miles, when the road crosses La Bonte creek, is the site of Camp Marshall on the left (east) side of the road.

Camp McGraw. 1857-58. Fort Thompson. This temporary field camp was built by a detachment of troops from Fort Kearny, Nebraska, who were surveying a road to Oregon that would miss the Utah Territory. There is a marker at the site on Wyoming 789 about 2 miles northeast of Lander.

Depot Fort McKinney. *Cantonment Reno.*

Fort McKinney. 1877-94. *Cantonment Reno, New Fort Reno.* The lineage of this permanent-type post is scrambled into its use of the Reno names in its early days, and the use of the McKinney name for awhile at the second Fort Reno. It was a direct descendent of Cantonment Reno when it was established in 1877 and for that reason the two alternate names were attached to the new fort until McKinney became official.

Seven double-story barracks, 14 officers quarters, and other frame buildings made up this large post by the time it was completed. In addition to Indian fighting, the garrison was critically involved in the nearby Johnson County War in 1892.

Several buildings remain, including part of the hospital and a stable, at what is now the Wyoming Soldiers and Sailors Home, 3 miles west of Buffalo and south of US 16.

Camp Medicine Butte. 1885-87. Two infantry companies were deployed to Evanston in September, 1885, immediately after the anti-Chinese riot in Rock Springs. The rioting having spread to the property of the Union Pacific railroad, the garrison was ordered to protect the U.S. mail aboard the

FORT LARAMIE
Officers' Quarters, about 1891

railroad trains. The issue was over the hiring of Chinese laborers in the mines, including those owned by the U.P., and the troops in Evanston were close enough to reinforce those at Camp Pilot Butte, if required. Evanston is on I-80 in the southwest corner of Wyoming; the camp was within the town.

Fort Nonsense. *Fort Bonneville.*

Camp O.O. Howard. 1885. This was a temporary Camp of Instruction 1 mile from Pine Bluffs, occupied by eight companies from the 4th, 7th, 9th and 21st U.S. Infantry from September 3 to 21, 1885. Pine Bluffs is 40 miles east of Cheyenne on I-80.

Fort Payne. 1855-59. *Camp Davis, Camp Clay, Camp of Platte Bridge.* John Baptiste Richard of Fort Bernard trading infamy, turned his talents to toll bridge operation from 1851 to 1865 with a small trading post on the side. At Richard's bridge seven miles east of Platte Bridge, two dozen soldiers guarded the bridge during the winter of 1855-56. During the Utah Expedition two companies of the 4th Artillery were garrisoned here in a small adobe fort. The location was nicknamed Camp or Fort Payne by the soldiers, possibly because of the "pain" they felt by having to spend the winter of 1855-56 at this desolate spot. From Casper, go east on I-25 to the next exit, Evansville. South of Evansville on the river are the archeological remains at the site of the blacksmith shop, cabin and trading post at which Fort Payne was located.

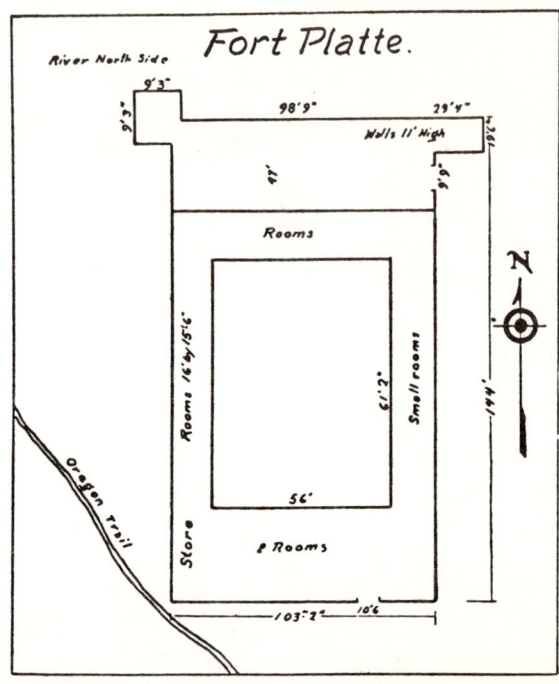

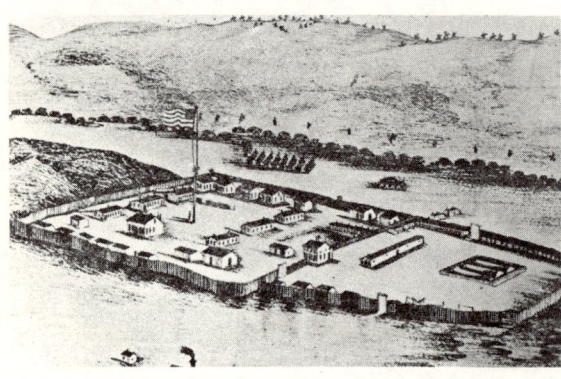

FORT PHIL KEARNY (1)

Fort Phil Kearny. 1866-68. *New Fort Reno.* Colonel Henry B. Carrington built this post to be the keystone of the Bozeman Trail, but most of the time it was under virtual siege. It was intended to be a replacement for Fort Reno, instead it was the headquarters of the Bozeman Trail forts of Reno, Phil Kearny, and C.F. Smith.

From this 42-building log palisade, Brevet Lieutenant Colonel William Fetterman led 66 men to their deaths in the Fetterman Massacre nearby. A year later, revenge of sorts was exacted near the fort in the Wagon Box Fight in which 1,137 Indians were killed or wounded when the troopers answered an Indian ambush with new breech-loading rifles.

There is a log cabin and partial stockade wall at the fort's site, neither of authentic origin. From Buffalo, go north on I-90 to Story, 14 miles. Turn west on Valley road, a short distance, to the Portugee Phillips marker (memorializing fort's guide who rode 236 freezing miles to warn of the Fetterman Massacre). On the left (south) of the road is a plateau, the site of Fort Phil Kearny.

Camp Pilot Butte. 1885-99. *Camp Rock Springs.* After a score of Chinese laborers were killed during a strike at the mines around Rock Springs on September 2, 1885, the government rushed troops into the city on September 5. Two companies remained behind when the bulk of the force was withdrawn a month later, and the post continued to be manned until the end of the century. The site of the camp still has several Army buildings in use. The parochial school building of the Church of Saints Cyril and Methodius was a barracks, since rebuilt and refaced with brick. Across the street are former barracks, now apartments, that retain their Army appearance. Rock Springs is on I-80 midway between Evanston and Rawlins.

Camp Platte. 1840-47. This was a camp site on the Platte river used by Oregon Trail trains. The ferry crossed the river at this point, later the location of Fort Caspar. For directions see Fort Caspar item.

Platte Bridge Station. *Fort Caspar.*
Camp at Platte Bridge. *Fort Payne.*
Portugese Houses. 1834-40. *Fort Antonio.* More legend than fact appears to have been written about this fur trade post established by Antonio Montero, a Portugese Mountain Man. A cluster of solidly built hewn log huts and a stout stockade 200 feet square provided protection in the hostile Powder River country.

According to tales the protection was enough for Montero to hold off an Indian siege for 40 days, but not enough for him to keep out 300 fellow trappers who descended on his area for the winter of 1836-37 and gradually absconded with everything movable.

From Casper, go north on I-25 to Kaycee, 72 miles. Turn right (east) at Kaycee on a gravel road that arrives in 11 miles at a marker for the site of the Portugese Houses.

Fort Reno. 1865-68. *Fort Connor, Old Fort Reno.* General P. Edward Connor led his troops from this new post on the Powder River Expedition of 1865.

Later when the Bozeman Trail was opened, Connor was to be abandoned but Colonel Henry B. Carrington decided to keep and remodel it after he led his 500-man expedition to it. The open post was enclosed with a stockade and renamed Reno on November 11, 1866. The post was abandoned when the Army withdrew from the Bozeman Trail in 1868.

From Casper, go north on I-25 to Kaycee, 72 miles. Turn right (east) at Kaycee on route 1002, going east 17 miles to Sussex schoolhouse. A quarter of a mile past the school, turn north on a gravel road. The Fort Reno (and Connor) site is on the right (east) side of the road, about 10 miles. A sign is beside the road, a marker in the pasture behind the marker and on a bluff overlooking the Powder.

Cantonment Reno. 1876-77. *New Fort Reno, Reno Station, Depot Fort McKinney.* Connected with the earlier Fort Reno only because of the similarity of names and geographical proximity, this post was three miles south and active only for nine months at the time of the Little Big Horn Sioux Expeditions of 1876. Consisting mainly of cottonwood log cabins and dugouts, it lasted into 1878 under the McKinney depot name while the garrison officially moved northward to newly established Fort McKinney. Only parched remains of building sites are left next to the Powder river. Although in the vicinity of Fort Reno, the unmarked privately owned site is impossible to find without local assistance.

Richards Post. *Fort Bernard.*
Camp Rock Springs. *Camp Pilot Butte.*
Detachment at Rocky Ridge. *Detachment at St. Mary's Station.*
Detachment at St. Mary's Station. 1862-65. *Detachment at Rocky Ridge.* As with the other stage station posts, the tiny garrison was responsible for the security of occupants and travelers, a task severly challenged on May 27, 1865, when the post was attacked. The occupants were able to hide on June 1 when the post was burned down and 400 yards of telegraph wire cut. From Lander, go south on US 287 for 38 miles to Sweetwater Station (a crossroads, not the historic Sweetwater Station further east). From this point where the highway crosses the Sweetwater river, St. Mary's Station is about 10 miles to the southwest on the banks of the Sweetwater.

Fort Sanders. 1866-82. *Fort John Buford.* This four-company post was arranged around a 400- by 235-foot parade ground when first built, then enlarged to a 600- by 500-foot parade when two more companies were assigned to the permanent garrison. Two buildings are left at the site, both of stone for obvious reasons: guardhouse and maga-

OLD FORT RENO *(1)*

zine. I-80 cuts right across the parade ground south of the city of Laramie. From the downtown area, take US 287 south 2 miles. A stone marker is at the junction of the highway and a dirt road which leads to the guardhouse ruins, 100 yards east. The magazine is on the other side of the interstate highway in front of the Laramie Country Club clubhouse.

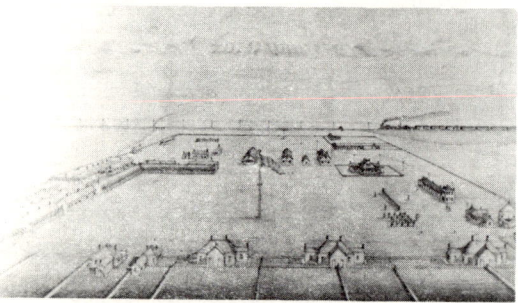

FORT SANDERS, 1875 (9)

Camp on Sage Creek. *Camp Hat Creek.*

Camp Scott. 1857-58. When the Utah Expedition arrived at Fort Bridger, they found that the Mormons had burned both it and Fort Supply to the south. The winter found the expedition quartered in rude huts, leantos, dugouts, and tents at Camp Scott 2 miles from the scorched remnants of Fort Bridger. Follow directions to Fort Bridger; turn south for 2 miles past the post to the approximate site of temporary Camp Scott.

Camp Sheridan. 1886-91. When the Army was directed to take over the security of Yellowstone Park, their first camp was this post, a number of wooden buildings sufficient for one cavalry troop. This was temporary, however, and when the permanent buildings of Fort Yellowstone were completed, Sheridan's site was abandoned. The camp site is south of the Yellowstone National Park headquarters at Mammoth Hot Springs and on the south side of Capitol Hill (in front of the hotel). The last remaining building of Camp Sheridan burned down in 1964.

Detachment at South Pass Station. 1862-66. Indians twice burned this station, giving it the name of "Burnt Ranch." Troops operated from it in protecting the entrance to South Pass.

In spring, 1865, three ex-Confederates serving here with the 11th Ohio Cavalry as Galvonized Yankees were shot to death for mutiny. Previously they had been in the Fort Laramie guardhouse for the same offense; their resistance to his authority was sufficient basis for the officer commanding the detachment to take the ultimate action warranted by field conditions.

From Lander, go south on US 287 for 6 miles to Wyoming 28. Turn right (southwest) for 24 miles to a left (southeast) turnoff to Atlantic City. A marker is on the highway 1.5 miles before the turnoff. At 2.3 miles turn right (southwest)

toward South Pass City. Pass through South Pass City ghost town and in about 3 miles turn left (southeast) for 6 miles to Burnt Ranch site on the Sweetwater river, site of South Pass Station. This is a rough, fair weather road; local directions should be obtained.

Camp Stambaugh. 1870-78. Two hewn log barracks, 80 by 32 feet, with an L, 48 by 20 feet, four sets of married soldiers quarters, four officers quarters, and other wooden buildings comprised this military protection for the miners in the Sweetwater district. By mid-1870 the drop in local population made the post unnecessary.

The modern site has traces of most of the buildings, now no more than mounds or outlines on the surface of the ground, and a historical marker nearby.

From Lander, go south on US 287 for 6 miles to Wyoming 28. Turn right (southwest) for 24 miles to a left (southeast) turnoff to Atlantic City. This is 1.5 miles past a marker to Atlantic City on the highway. At 2.3 miles turn left (southeast) to Atlantic City. Out of Atlantic City, follow the Micro Wave Tower road about 2.8 miles to a dirt track entering from the left (north). Follow this dirt track across the plain to the marker; the site and the surface traces are beyong the marker. This is a rough, fair weather road; local inquiry is essential.

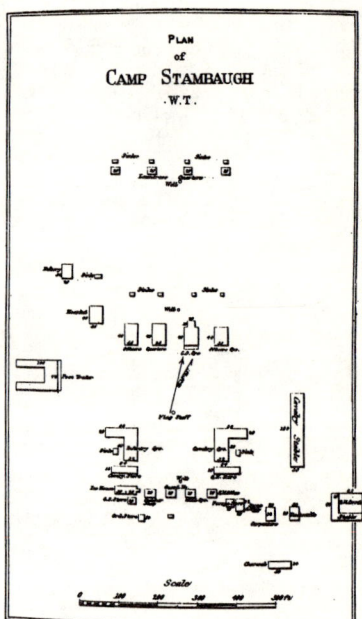

Camp Stillings. 1862-66. *Detachment at Sweetwater Station.* Forty Arapahoes attacked this garrison in June, 1865, but were beaten off by the 14 soldiers who lost one trooper killed, one wounded. Only excavations are left where the buildings once stood. From Casper, go west on Wyoming 220 about 55 miles to the crossing of the Sweetwater river. Turn left (west) 1 mile to the site of Sweetwater Station and Camp Stillings.

Fort Supply. 1853-57. Not really a military fort, this was the site to which the Mormons sent representatives in an attempt to establish a farm to support emigrating Mormons. A high stockade surrounded the cluster of buildings, all of which was burned when the Mormons abandoned the area upon the approach of the Utah Expedition. This post represents the first effort at farming in Wyoming, as the plaque at the site points out. Follow directions to Fort Bridger, then turn south through Mountain View to Robertson, 12 miles. One mile west of Robertson is a marker at the site of Fort Supply.

Detachment at Sweetwater Station. *Camp Stillings.*

Fort at TA Ranch. 1892. The Johnson County war between the cattle barons and the independent cattle ranches reached its highpoint here when forces of the barons started to storm this tiny defensive work at the TA ranch. The "fort" was on top of a knoll, deep trenches enclosing a space 12 to 14 feet square surrounded by a wall of heavy timbers with firing portholes. The permanent buildings of the ranch were logged up for protection, the main ranchhouse also has a series of trenches around it. The attackers were behind a moving barricade but before it had a chance to challenge the defenders, regular Army troopers from Fort McKinney arrived to suspend the conflict. The TA ranch location is south of Buffalo about 13 miles on the east side of I-25 (US 87).

Fort Thompson. *Camp McGraw.*

Detachment at Three Crossings. 1862-66. It was a true cavalry finish when the troopers stationed at this stage station were surrounded by an estimated 300 to 500 Indians on May 20, 1865. Troops from other stations arrived just in time to rescue the garrison. This site also was the headquarters post for Company I, 3d U.S. Volunteers, Captain A. Smith Lybe, which provided the details for three other posts. This company marched farther and had more casualties than any other company in the regiment. It was known also as Galvanized Yankees because of their former Confederate allegiance. From Lander, go south on US 287 for 58 miles to where the highway nears the site of the station on the left (north) side.

Camp Walbach. 1858-59. Protecting the dangerous crossing through Cheyenne Pass was the mission of Camp Walbach, described in 1866 by a passing traveler: "A small military post that had been entirely destroyed by Indians years before, vividly reminding us of the deadly foe, not far away." Whether destroyed by Indians or firewood-hungry travelers, the questions of who destroyed the post is academic since anything untended was fair game at the time. The camp was manned by two companies of the 4th U.S. Artillery between September, 1858, and April, 1859. From Cheyenne, take I-25 north 7 miles to the exit to Federal. Take this exit to the west about 16 miles to Federal. Inquire here for directions to Cheyenne Pass and Camp Walbach, about 4 miles to the west.

Fort Washakie. 1871-1909. *Camp Brown.* A rarity, this fort was named after an Indian chief, Shoshone leader Chief Washakie who was also the leading resident at this permanent stone garrison in the Wind River area. A number of buildings still are in use at the Wind River Indian Agency. From Lander, go northwest on US 287 for 17 miles to the town of Fort Washakie. The fort buildings can be seen from the road on the left (west) side. The post cemetery, with the grave of Chief Washakie in it, is about 4 miles south of the former post.

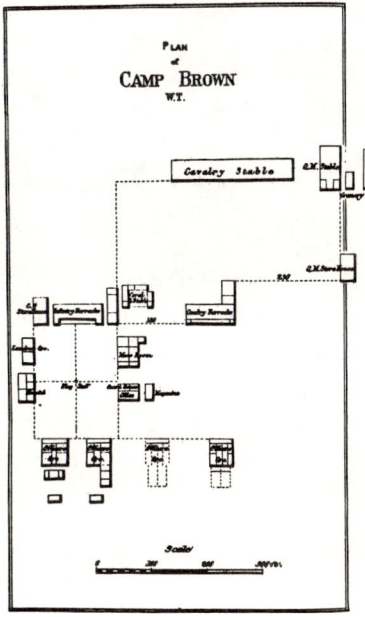

Fort William. *Fort Laramie.*

Camp Winfield. 1857. *Camp on Ham's Fork.* Upon arriving at Fort Bridger and finding it burned, Colonel E.B. Alexander took his forces about 30 miles to the northwest and made camp at this point. Much indecision and correspondence with Brigham Young followed before the troops left the camp on a northwesterly route, intending to descend upon Salt Lake City from the north.

Six days later the seven mile long column ground to a halt 35 miles from Winfield. It turned around and returned to Winfield after another 16 days, having spent from October 11 to November 2, 1857, making a great circle without accomplishing anything. By this time, General Albert S. Johnson had arrived and ordered the column to begin moving on November 6 for the site of what became Camp Scott, 35 miles to the south.

From Fort Bridger on I-80, go northeast on I-80 to US 30W. Turn to the northwest on US 30W through Granger. Twenty miles from the interstate — which is about 15 from the junction of Ham's Fork and Black's Fork — is the site of Camp Winfield.

Fort Yellowstone. 1892-1918. Upwards of four troops of cavalry and one machine gun platoon garrisoned this permanent post designed for the protection of Yellowstone Park before the organization of the National Park Service. This post succeeded and was just north of Camp Sheridan. The buildings are now used by the administration staff of the National Park Service at Yellowstone in the Mammoth Hot Springs area.

Chief Washakie talking to a group at Fort Washakie *(1)*

SOUTH DAKOTA

FORT PIERRE *(1)*

The Missouri river has left its mark on the sites of the river forts of South Dakota, the Oahe reservoir having inundated many of those of central and northern South Dakota, the Fort Randall reservoir doing likewise for those in the southern part of the state.

Both trading and military posts have been inundated by the Missouri, but the pre-flooding archeological work of the State Historical Society, the National Park Service, and the Smithsonian Institution preserved much of the history and artifacts of the sites. The reports of these projects aided in the compilation of this section.

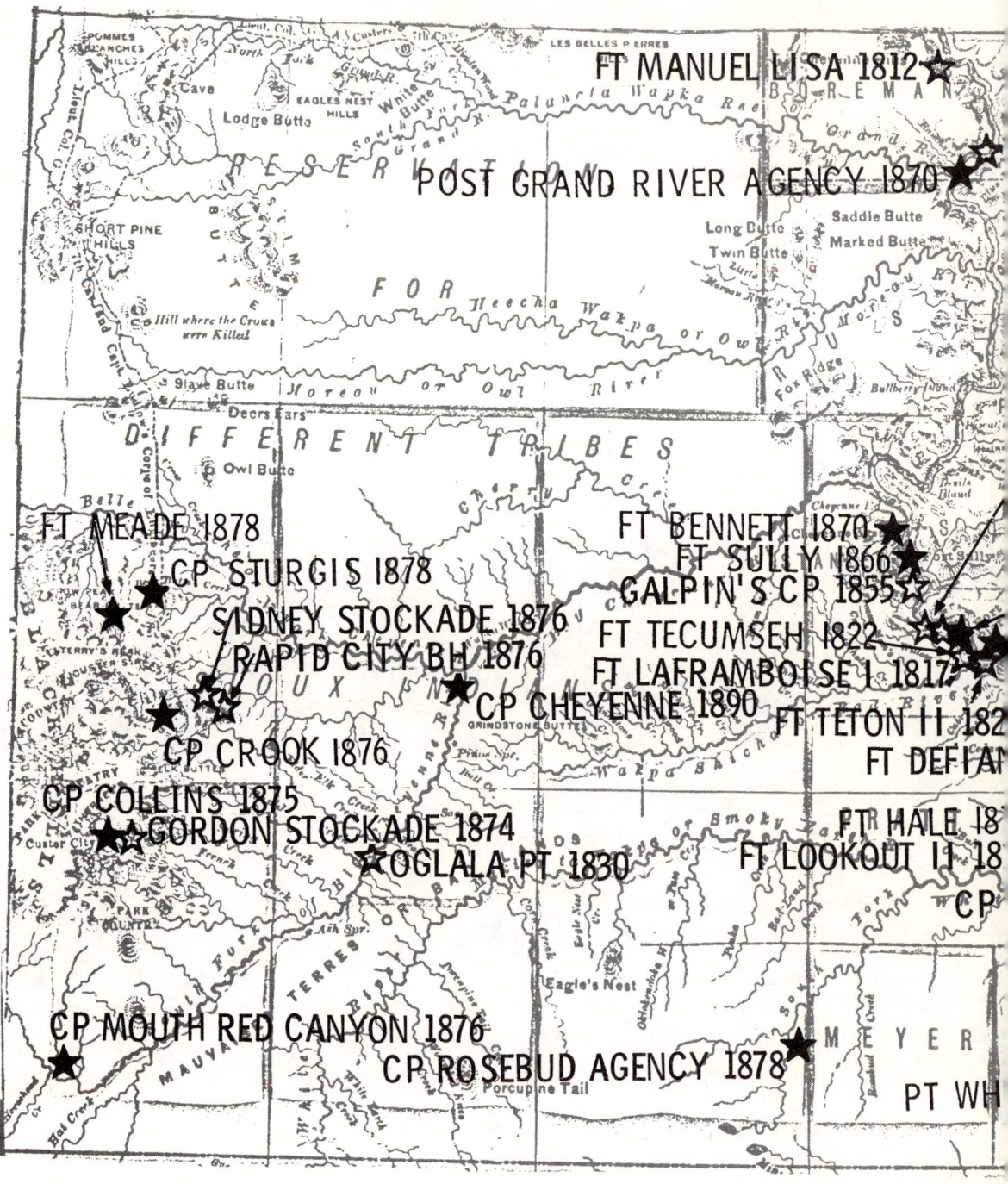

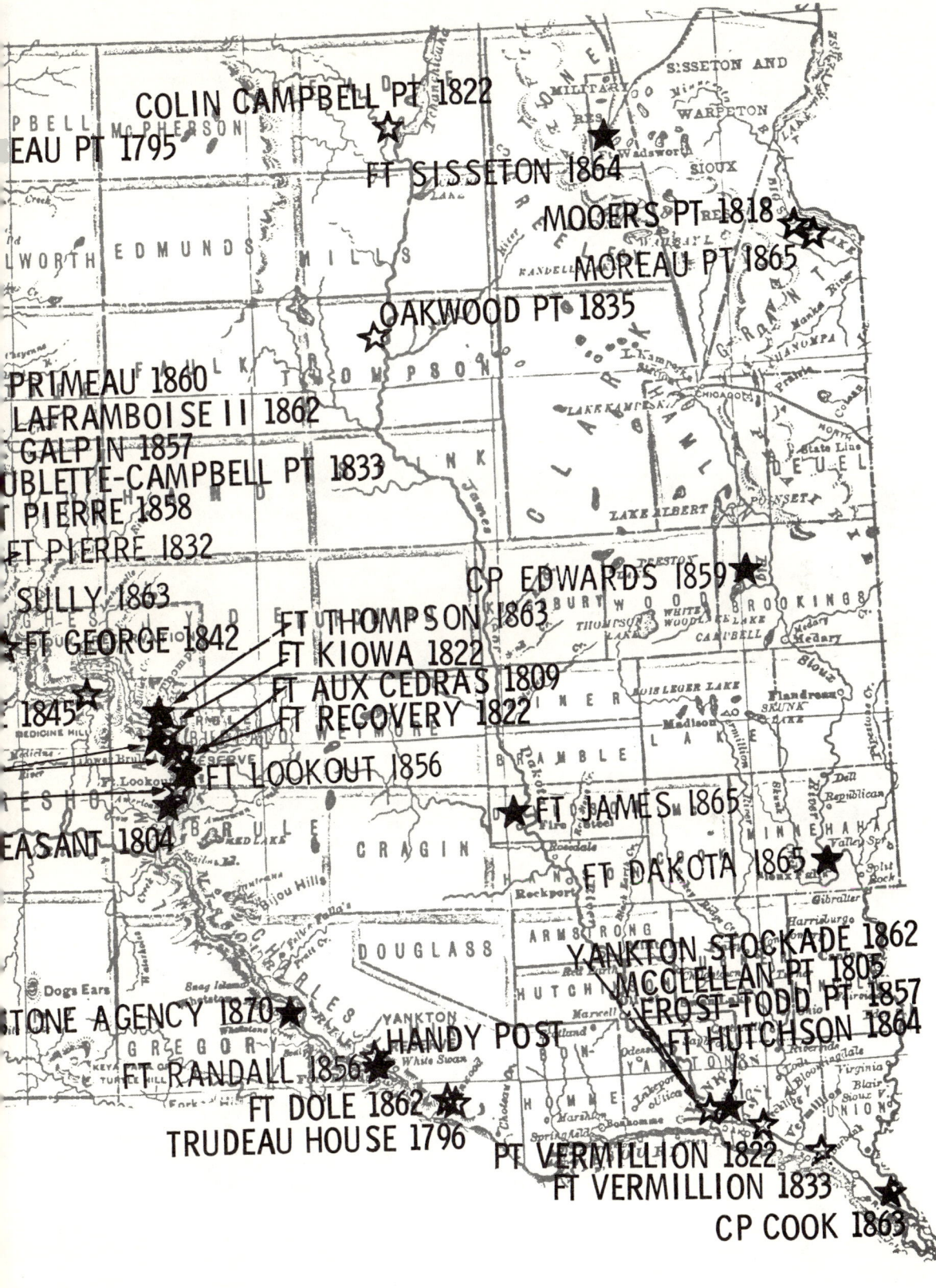

Fort aux Cedras. 1809-1822. *Loisell's Post.* This Missouri Fur Company trading post was on American Island—called Cedar Island at the time—in the Missouri river. An important post during its active years, the fort was burned in 1822. Supposedly it was replaced by Fort Recovery, but Smithsonian Institution research indicates that Fort Recovery, although it may have inherited aux Cedras' market, did not occupy the same site. American Island was in the Missouri river about .8 mile south of Chamberlain; it has been inundated by the Fort Randall reservoir.

Fort Bartlett. *Old Fort Sully.*

Fort Bennett. 1870-91. *Post at Cheyenne Agency.* A semi-stockade with its open sides on the Missouri river, to the east, and Agency creek to the north, Fort Bennett was intermixed with buildings of the Indian Agency. It occupied three sites, moving as river erosion threatened the various locations, but all within 300 yards of each other. Usually a one-company post, Fort Bennett had a nine-company garrison during the aftermath of the Custer massacre in 1876. The site has the remnants of three buildings about 35 miles northwest of Pierre; although inundated by the Oahe Reservoir, it is exposed periodically during low water in the reservoir.

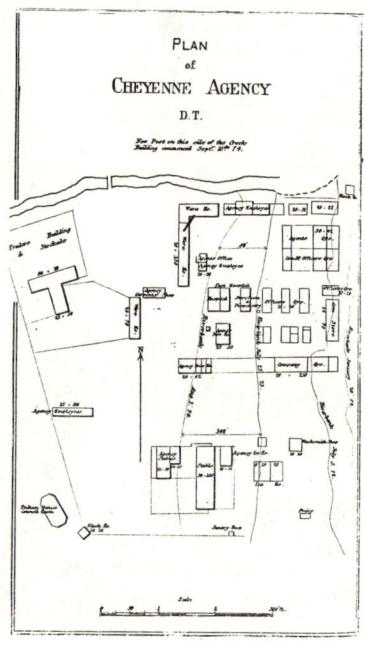

Fort Bouis. *Fort Defiance.*
Fort Brasseaux. *Fort Recovery.*
Fort Brookings. *Fort Dakota.*
Cedar Fort. *Fort Recovery.*

Camp Cheyenne. 1890-91. *Camp of Observation.* Five companies, two Hotchkiss guns, and a detachment of Indian Scouts were stationed at this field camp near the Forks of the Cheyenne rivers during the days leading up to the Messiah War. In addition to the obvious mission described by its alternate and initial name, Camp of Observation, the post was a base of operations for what culminated in the Battle or Massacre of Wounded Knee. From Rapid City, go east on I-90 for 35 miles to Wasta exit. Take a secondary road north from Wasta about 20 miles to Elm Springs. Turn right (east) on a gravel road that passes the approximate site of Camp Cheyenne in 25 miles on the right between the road and the Cheyenne river.

Post at Cheyenne Agency. *Fort Bennett.*

Fort Chouteau. *Fort Pierre.*

Colin Campbell Trading Post. 1822-28. This was a stockaded post built for the Hudson Bay Company. From Aberdeen, go north on US 281 for 17 miles to South Dakota 10. Turn left (west) for 1.2 miles on South Dakota 10 to a dirt road. Turn right (north) for 1.6 miles to a monument on the site of the trading post.

Camp Collier. *Camp at the Mouth of Red Canyon.*

Camp Collins. 1875-76. The Log Cabin Museum in Way City Park on Main street in Custer is reputed to have been built by the Army as the headquarters for a three-company detachment stationed here to keep the miners out of the Black Hills. The cabin is 16 by 20 feet with loopholes. Custer is in the Black Hills National Forest at the junction of US routes 16 and 385.

Camp Cook. 1863. Iowa Volunteers camped at this site for a week in May, 1863. At the time it was located 6 miles west of Sioux City, Iowa, on the west bank of the Big Sioux river. The modern site is about 2 miles north of North Sioux City to the east of US 77.

Camp Crook. 1876. General George Crook established this as his headquarters during the Army attempts to oust prospectors from the Black Hills. From Deadwood, go south on US 85 for 2 miles to US 385. Turn left onto US 385 for 27.3 miles to Pactola recreation area, general site of Camp Crook. The camp has been inundated by the Pactola reservoir.

Post at Crow Creek Agency. *Fort Thompson.*

Fort Dakota. 1865-69. *Fort Brookings.* The dozen buildings of Fort Dakota helped bring Indian-wary settlers back to the Sioux Falls area. The fort was so successful in this mission that it was soon all but overrun by homesteaders who wanted to file claims on the military reservation.

FORT DAKOTA, 1866 *(I)*

There are no remains, the site having been swallowed by the city of Sioux Falls, but a marker to the fort is on the exterior wall of the Hollywood theater, 218 North Philips avenue in downtown Sioux Falls. Across the street was the fort site: the barracks was on Philips with the south end of the barracks 125 feet north of 8th street.

Fort Defiance. 1845-49. *Fort Bouis.* When two ex-clerks of the American Fur Company set up a trading post in competition with their former employer, their defiance served as an appropriate name for the fort. A less antagonistic name was Bouis, after a member of the firm. The approximate site was on the west bank of the Missouri near the mouth of Medicine Creek, probably inundated by Lake Sharpe. The general area can be reached from Chamberlain via I-90 west to Reliance. Turn north at Reliance on South Dakota 47W for 8 miles to a fork with a secondary road that heads straight north (47W turns to the right—east—at this point). Go north on this road through Lower Brule; about 7 miles west of Lower Brule the road crosses Medicine Creek and the approximate site of Fort Defiance.

Fort Des Roche. *Fort James.*

Dickson's Post. *Post Vermillion.*

Fort Dole. 1862. Yankton Indian Agent Walter Burleigh took no chances when he heard that the Upper Sioux were on the warpath. He built this double-storied octagonal blockhouse, 26 feet in diameter, of 22-inch thick timbers, loopholed for muskets and defenses by a six-pounder and two three-pounders. The Dole came not from the usual activity with the Indians but from the last name of the Commissioner of Indian Affairs. From Pickstown, next to the Fort Randall Dam, go east on South Dakota 46 for 5 miles to an intersection; turn right (south) through Marty, 5 miles, to Greenwood, an additional 5 miles, once the Yankton Indian Agency and the site of Fort Dole.

Camp Edwards. 1859; 1862; 1864-65. *Cantonment Oakwood, Camp near Preston Lake.* Earthen breastworks five feet high about 100 feet square were erected around a single log hut at this post, occupied in 1859, again after the Sioux Outbreak, and finally by Minnesota Volunteer Cavalry at the end of the Civil War. The outline of the breastworks is in the Oakwood Lakes State Park.

From Brookings, go north on US 77 for 8 miles to County 30. Turn left (west) through Bruce, 4 miles, and after 3 more miles turn right (north) into the Oakwood Lakes.

Follow marked road north between three lakes to the site of Camp Edwards, indentified by a marker intitled "Breastworks;" the Camp Edwards name had been lost until unearthed in National Archives research of regimental records.

Camp near Firesteel Creek. *Fort James.*

French Post. *Fort Lookout II.*

Frost-Todd Trading Post. 1857-61. The firm of Frost and Todd, the latter a cousin of Mary Todd Lincoln who made full use of his connections with the White House, monopolized trade with the Indians and travelers at this post near the Yankton Agency. From Yankton, go east on South Dakota 50 for 3.7 miles to the James river; the site of the trading post is on the left (north) side of the road at the river crossing. The post was moved to Yankton, at the southwest corner of 2d and Walnut streets.

Galpin's Camp. 1855-57. After Charles E. Galpin, serving as the agent of the Chouteau family, engineered the sale of the first Fort Pierre to the Army, he and other former employees from Fort Pierre moved to this site. It appears that they conducted trading operations with the Indians during the winters of 1855-56 and 1856-57 before moving south and establishing Fort Galpin. The site is now inundated on the western edge of the Oahe Reservior about 14 miles directly northwest of Pierre.

Fort Galpin. 1857-59. This trading post was 125 feet square, similar to the first Fort Pierre but without any blockhouses. Two-thirds of the enclosure was a picket stockade and houses closed the remainder. From Pierre, go west across the Missouri river on US 14. Turn right (north) onto South Dakota 514, the Oahe Dam road, for 4 miles. The approximate site of Fort Galpin is to the right (east) of the road, next to the river and just north of the site of the second Fort Pierre.

Fort George. 1842-45. This Fox, Livingston and Company trading post was a stockade 155 by 165 feet with projecting blockhouses at two opposing corners. Smithsonian archaeological excavations in recent years determined that this post was unusual in that the buildings were separated from the stockade walls by an alleyway rather than built against the walls. Liquor and violent competition proved the end of Fort George and it was burned by Indians in the pay of rival traders.

Gordon Stockade. 1874-75. The gold fields of the Black Hills were the targets of the 28 occupants of this 80-foot square, 10-foot high stockade. Their prospecting was successful, even finding it when they dug an eight-foot deep hole in one corner of the stockade. Government policy prohibited incursions in the Black Hills at the time and the U.S. Cavalry was called out to escort the prospectors out of the area after the winter. A replica of the stockade has been erected on the north (left) side of US 16 alternate 3 miles east of Custer.

Post at Grand River Agency. 1870-75. This two-company fort adjacent to the Grand River Agency included 20-some buildings, all of log and mud-chinked construction with plank-and-mud roofs. The nearby Missouri river periodically flooded the post and finally forced the removal of the agency and garrison to Fort Yates. The site finally surrendered to the river with the Oahe Dam project and is inundated about 5 miles north of Mobridge.

Fort Hale. 1870-84. *Post at Lower Brule Agency, Fort Lower Brule.* Protecting the Lower Brule Indian Agency was only part of the mission of Fort Hale. The garrison frequently had to share its commanding officer with the agency because

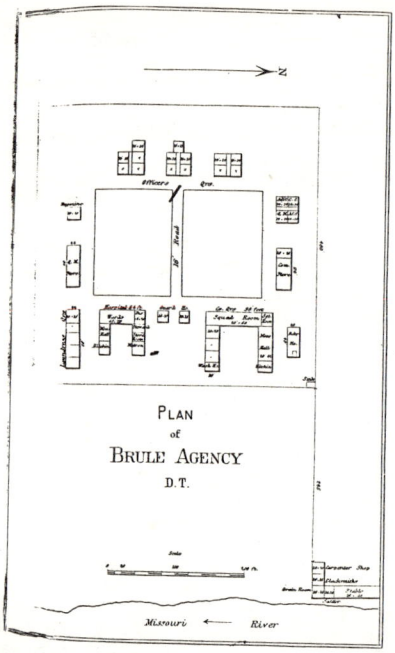

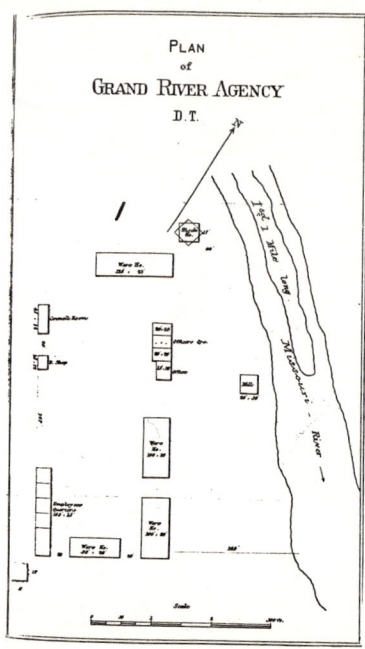

the Army commander served also as the acting Indian Agent. With the commanding officer also serving as the agent at the Crow Creek Agency, this arrangement was less than satisfactory.

The post was originally located near Fort Lookout IV on the west bank of the Missouri but soon was moved 15 miles to the north. Bricks from post buildings were used in the construction of the Chamberlain hotel in Chamberlain; the rest of the post has fallen vicitm to the Missouri river. The final traces, cellar depressions and gravelled walkways, were washed away by the flood of 1952.

From Chamberlain the approximate site can be reached by going north on South Dakota 47 to South Dakota 34, then left on 34 through Fort Thompson town and across the river on the Big Bend dam. The site was south of the dam on the west side of the river.

The site of the first Fort Hale (1870, usually called Fort Lower Brule) is on the west bank of the Missouri almost directly opposite Chamberlain and to the north of US 16 (about 5 miles north of the Oacoma exit on I-90). There are low hammocks on the second terrace above the river which are reputed to be the site.

Handy's Post. On the site of Fort Randall, little is known of this early-day trading post.

Fort Hutchson. 1864. This short-time Civil War Volunteer post was east of Yankton. The approximate site is shown on a state marker at the northeast corner of the intersection of I-29 and South Dakota 50 about 8 miles east of Vermillion.

Fort James. 1865-67. *Camp near Firesteel Creek, Fort Larouche, Fort Des Roche.* This was a quadrangle made of stone and hewn logs that was intended to garrison troops for stage coach protection. It was built by professional stone masons and would have lasted after abandonment except that its materials were of value in other construction. From Sioux Falls, go west on I-90 about 55 miles to Alexandria turnoff. Go west on county road from the north end of town. At 5 miles, turn south for 5.3 miles, then turn east for 2 miles. At bluff, drop down to the first right. Pass through the Hutterite Colony and just past the colony is the fort site on the left side (east) of the road between the road and the James river.

Fort Kiowa. 1822-25. *Fort Lookout I.* Four posts used the Lookout name in South Dakota and the decision to attach the first one with Fort Lookout is based on Smithsonian Institution research in 1950. This was a fur trading fort built by the American Fur Company, consisting of a range of log buildings containing four rooms, a log house and a storehouse forming a right angle, leaving a space of some thirty feet, according to a report quoted by Chittenden. Pickets surrounded the post, including a blockhouse at the south corner and a wooden tower at the north; each side was reported to measure 140 feet. The approximate site is believed to be inundated by the Missouri river directly east of Fort Hale II.

LaBarge's Post. *Fort Lookout II.*

Fort LaFramboise I. 1817-20. *Fort Teton I.* Joesph LaFramboise built this trading post of dead logs which he rescued from the nearby Missouri. The post operated for only a couple of seasons and then was abandoned. The approximate site is

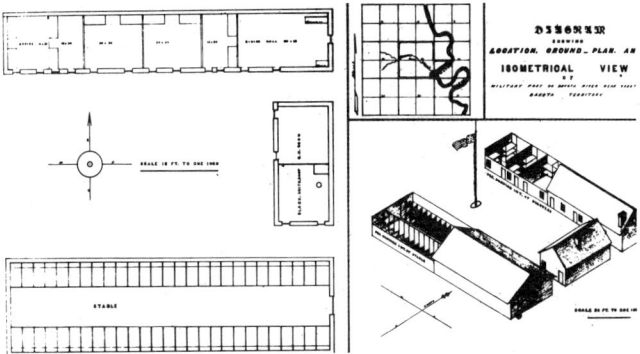

FORT JAMES (1)

the same as the modern city of Fort Pierre, across the Missouri river from Pierre.

Fort LaFramboise II. 1862-63. Built by LaBarge, Harkness and Company to oppose the Fort Pierre trading, this post was abandoned within a year. It had no stockade, the buildings themselves making up an enclosure. The approximate site is at the western side of the Oahe Dam off of South Dakota 514 northwest of Pierre.

Fort Larouche. *Fort James.*

Loisell's Post. *Fort aux Cedres.*

Fort Lookout I. *Fort Kiowa.*

Fort Lookout II. 1831-51. *French Post, LaBarge's Post.* Under these three names this site was occupied periodically during the second quarter of the nineteenth century, going under the French Post name in 1833 and with that of the famous St. Louis trading and steamboat LaBarge family between 1840 and 1851.

Smithsonian Institution archeological research in 1950 and 1951 apparently located the site of a rectangular trading post dating from the earlier portion of these dates. This appeared to have fallen into partial ruin and then been burned after which a second post was built on the same site that suffered the same fate. Both appeared to measure about 70 by 20 feet, consisting of a single building each time and without any evidence of a stockade.

Follow the directions to Fort Lookout IV; at the entrance to the Lower Brule Reservation, where a state marker indentifies the historical sites in the area, the location of Fort Lookout II is to the south about 300 yards in the pastureland along the Missouri river.

Fort Lookout III. The dates of this trading post are as uncertain as whether it is different from Fort Lookout I (Fort Kiowa). It may be that this post, definitely recorded as being in existence in 1833, is the same as that of Fort Kiowa and that is the conclusion accepted by Smithsonian Institution research. All of this is academic, anyway, in that the site is under the Missouri river to the east of the Fort Hale location, making definite determination impossible as to whether it was the same or different from the other Lookout post.

Fort Lookout IV. 1856-57. The only military post of the four—or three, depending upon the authority—Lookouts, this was intended to be an elaborate one. The parade ground was a quarter of a mile long, flanked by barracks and closed by three officers quarters at one end and the guardhouse and headquarters building at the other. After only a season of operation, the post was abandoned and much of its construction moved downriver to Fort Randall.

From Chamberlain, go west on I-90 to Reliance, about 20 miles. Turn right (north) at Reliance to first right turn, 1 mile. Take this dirt road for 11 miles to where it turns south at bluffs overlooking Missouri River. Take primitive road that drops from here to river valley. At end of the road is a marker for the Lower Brule Indian reservation.

The Fort Lookout IV site is to the northeast of the marker, within the reservation and next to the river; slight rises in the ground may be apparent. The site of Fort Lookout II is to the south, 300 yards.

Fort Lower Brule. *Fort Hale.*

Fort Manuel Lisa. 1812-13. Dakota and Wyoming historians disagree on whether Indian girl Sakakawea died and was buried here or at Fort Washakie, Wyoming, with theories as to the ultimate fate of Lewis and Clark's guide drawn upon state loyalties. Regardless, this trading post was the home of Sakakawea's husband during its active period, including December 20, 1812, when at least one contemporary journal notes that his Indian wife died. Whether Sakakawea was the wife in question continues to be debated.

The fort was a stockaded post that supposedly was burned after the traders abandoned when 15 of their number were killed by the Sioux. The stockade was reconstructed by the Civilian Conservation Corps in 1930's. Its deteriorated condition was unchecked by the time the Oahe Reservoir was established because the fort stood at the precise high water point of the reservoir, about 30 miles north of Mobridge to the east of Kenel. There is a state marker on the road that provides directions to the site, also identified by another state marker.

FORT MEADE, 1888 *(30)*

McClellan Post. 1805-06. The second trading post on the Upper Missouri, this fort was in operation when Lewis and Clark came downriver in 1806. The site is east of Yankton on South Dakota 50.

Fort Meade. 1878-1944. *Camp Ruhlen.* Meade was the first Army post to play the Star Spangled Banner at retreat ceremonies, later a custom adopted by all Army posts even before it became the national anthem. Many Army buildings are now used by the Veterans Administration Hospital and the post cemetery is now a picturesque national cemetery. From Sturgis, take South Dakota 34 for 2 miles east to a right (south) turn into the Fort Meade VA Hospital.

Mooer's Trading Post. 1818-24. There is nothing but a marker at the site of this early-day trading center which was visited by the Stephen Long expedition in 1823. From Milbank (at the junction of US 77 and 12), go northeast 11 miles on US 12 to Big Stone City. Take South Dakota 15Y north from Big Stone City for 11 miles to Hartford Beach State Park, the site of the trading post.

Moreau-Robar Post. 1865. This trading post was established by Moses Moreau and Solomon Robar for the Missouri river trade. The site is 2 miles south of the Mooers' Trading Post, at Linden Beach.

Camp at the Mouth of Red Canyon. 1876-77. *Camp Collier.* A stockaded affair 125 feet square, bastions at the northeast and southwest corners, this short-term post watched over the Black Hills stage that passed through Red Canyon. Indentations in the pasture and the rotted stump of a stockade picket are at the site. From Edgemont, in southwestern South Dakota, go north on US 18 for 1 mile. Turn left at Red Canyon road for 5 miles to a point just before entering Red Canyon. The camp site is to the right of the road about 50 yards, on privately owned property.

Oakwood Trading Post. 1835-51. Joseph R. Brown established this fort but its manager, Pierre LeBlanc, did not fare too well: he was murdered by an unhappy customer. A DAR marker and a worn doorsill next to it are the only traces of this post in Rondell Park (named after a later operator of the post and a Dakota pioneer). From Aberdeen, go south on US 281 for 16.7 miles to the intersection with a secondary road. Right on this road is the hamlet of Mansfield but left 8 miles is Rondell Park and the site of the Oakwood Trading Post.

Oakwood Cantonment. *Camp Edwards.*

Oglala Post. 1830-34. The American Fur Company erected this fur trading post and it was here that Thomas L. Sarpy was blown up when a candle ignited a barrel of gunpowder. Although part of the fort was destroyed in the explosion, it continued in operation for a few years until the Indians moved to the vicinity of Fort Laramie. The site supposedly was used as a headquarters camp by the 6th Cavalry during the Messiah War in 1890. The site is the same as the ghost town of Link at the mouth of Rapid creek. From Rapid City, go southeast on South Dakota 40 about 35 miles to the Cheyenne river. At this point the site of Oglala Post can be seen to the north on the north side of Rapid creek.

Camp of Observation. *Camp Cheyenne.*

Pawnee House. *Trudeau's House.*

Fort Pierre. 1832-63. *Fort Chouteau.* One of the most important trading posts on the Upper Missouri, this stockaded fort was also the headquarters of General Harney's troops in 1855-57.

When the troops arrived, the American Fur Company employees moved upstream and built a second Fort Pierre, later using timbers from the first fort after the troops left. This second Fort Pierre operated from 1858 to 1863.

After the troops gave up the rotted and in-

FORT PIERRE
From a drawing by Alexander H. Murray *(1)*

adequate first Fort Pierre, it was usually called Fort Chouteau or Fort Pierre Chouteau to differentiate it from the newer post. From Pierre go west on US 14 across the Missouri river. Do not turn into the town of Fort Pierre; instead continue west on US 14 for 1 mile beyond US 83 to a righthand turn onto South Dakota 514. Go north 1.3 miles where the fort site marker, on a large boulder, is .3 mile east (right) of the road in a field at the end of a dirt road that crosses private property. This is the site of the first Fort Pierre.

The second Pierre site is 3 miles to the north and unmarked.

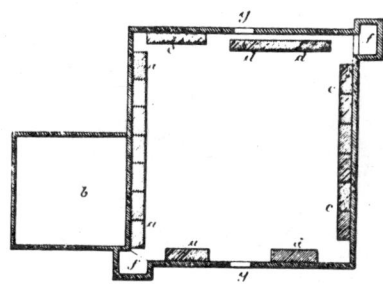

GROUND PLAN OF A TYPICAL TRADING POST
(Fort Pierre)
From a drawing by Maximilian

ff. Two-story block-houses. Upper story adapted for use of small arms; lower story for cannon.
gg. Front and back of quadrangle 114 paces in length; other sides 108 paces; inner area 87 by 87 paces.
dd. One-story residence of bourgeois of post.
e. Office and residence of clerk.
aaaa. Residence of other clerks, interpreter, engagés, and their families.
cc. Stores.
gg. Entrance doors to fort.
b. Garden.

Pilcher's Post. *Fort Recovery.*

Camp Pleasant. 1804, 1806. Lewis and Clark used this campsite for two days in September, 1804, on their way to Oregon and stopped overnight in August, 1806, on the way back. It is one of many Lewis and Clark sites along the Missouri river but one of the few marked ones. The marker is on US 16 about 3 miles west of Chamberlain and just north of Oacamoa.

Camp near Preston Lake. *Camp Edwards.*

Fort Primeau. 1860. This was a temporary trading house of LaBarge, Harkness and Company, named for Charles Primeau, one of the firm's partners. It apparently operated in the late 1850's or early 1860's. The site is under the Missouri near the western bank about 1 mile north of the Oahe Dam.

Fort Randall. 1856-84. First in the chain of military forts along the Upper Missouri, Randall was the base of operations for most of the Sioux Expeditions and, during the Civil War, one of the major forts garrisoned by ex-Confederates, so-called "Galvanized Yankees." Remnants of the stone chapel and many marked foundation and cellar outlines are at the site, now a park below the western corner of the Fort Randall Dam. Follow the signs from Pickstown across the dam.

Rapid City Blockhouse. 1876. White settlers who sifted through Army blockades to prospect

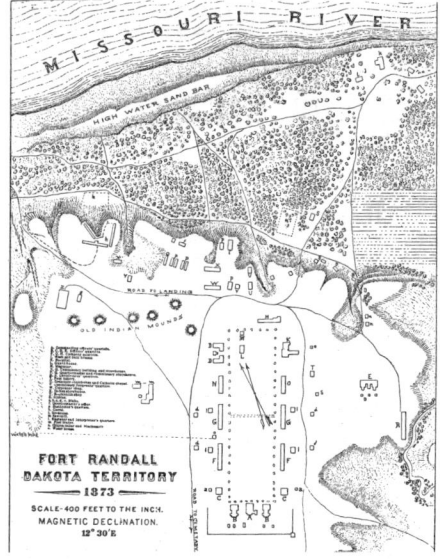

37

FORT RANDALL (1)

the Black Hills had to deal with official government opposition and deadly Indian hostility. After 180 of the original 200 founders of Rapid City abandoned the settlement in the wake of Indian ambushes on the outskirts, the remaining holdouts built this double-story, 30-foot square blockhouse for protection. The group was virtually besieged here for most of September, 1876, until word came through that the government had ceased its opposition to settlement. The site is at the intersection of 5th and Rapid streets in Rapid City.

Fort Recovery. 1822-1830. *Cedar Fort, Pilcher's Post, Fort Brasseaux.* Both the dates and additional names for this important trading post are confusing and uncertain, as uncertain as the site. Most authorities consider that the post was built on the site of Fort aux Cedres when that American Island post burned. Modern Smithsonian Institution archaeological surveys have concluded that it was on the right (west) bank of the Missouri river about even with the southern tip of the island. The matter is academic now: both sides, island or otherwise, have been inundated by the Fort Randall dam project. The fort's most important activity took place in 1823 when it was the base for the Ashley-Leavenworth Expedition against the Arikara Indians. The site would be about 1 mile south of Chamberlain on the western bank of the Missouri river, prior to flooding.

Military Camp at Rosebud Agency. 1878-91. Sixteen companies of soldiers were stationed at this post at the height of its strength to both control and protect the Brule Sioux moved here from the former Spotted Tail Agency in Nebraska.

The cemetery includes the grave of Chief Spotted Tail, marked by a white shaft. The hill above the agency is called Soldier Hill in memory of the rifle pits dug there by the troopers during the Messiah Craze.

The reservation headquarters is at Rosebud which can be reached from the Murdo exit on I-90, then heading south on US 83 for 43 miles to US 18, turning right (west) for 3 miles then heading southwest on a gravel road for 6 miles to Rosebud. The deep cut at the edge of the Agency, nicknamed the "Big Hole," is obvious in contemporary photographs. The cemetery is on the hill to the right.

Camp Ruhlen. This was the first name of the site of Fort Meade, attached to it during the time that Lieutenant George Ruhlen was the quartermaster in charge of the construction of what became Fort Meade.

Fort Sisseton. 1864-90. *Fort Wadsworth.* The peace of the Kettle Lake region of eastern Dakota was kept by this breastwork-surrounded post. The remains of most of the buildings are now at this state park site. From Watertown, take US 81 north to Sisseton, 57 miles. At Sisseton, turn left on South Dakota 10 and go west 25 miles. At gravel road six miles west of Lake City, turn left, go nine miles south. Fort buildings are on the right.

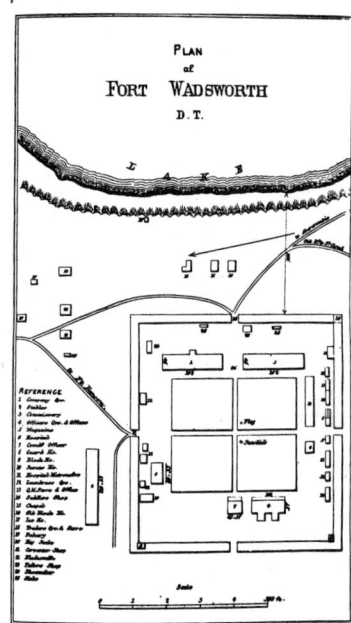

Sidney Stockade. 1876. Guarding the Sidney-Deadwood, Pierre-Deadwood stage lines were stage stations such as this reproduction on the site of the first station north of Rapid City. From Rapid City take I-90 northwest to the exit after Blackhawk. The reproduction is near the highway, right (north) side.

Camp Sturgis. 1878. Here in the shadows of Bear Butte, majestic mountain that dominates

FORT SISSETON
From *Northwest Magazine*, March, 1886

the Sturgis area, troopers constructing nearby Fort Meade encamped. A straight row of a dozen 'A' tents plus officers and supply tents were along Sturgis creek. Custer supposedly camped at the same site in 1874 and a picture of a camp below Bear Butte is variously used to illustrate either camp although local historians insist that it dates from 1878.

Sublette and Campbell's Post. 1833-34. Intended to operate in competition with Fort Pierre, the degree of success of this trading post can be hinted by the fact that it was sold to the American Fur Company after only a year. The site is immediately south of the Fort Tecumseh site.

Old Fort Sully. 1863-66. *Fort Bartlett.* At an unhealthy location of the Missouri river lowlands, this 270-foot square stockade served mainly during the aftermaths of the various Sioux campaigns of General Alfred Sully. A stone post is the only evidence on the site. From Pierre, go east on South Dakota 34 for 4 miles to where the marker is on the side of highway.

New Fort Sully. 1866-94. The direct descendent of Old Fort Sully, this was probably the most active on the Missouri in terms of operations fielded from it during its active years. It started as a stockaded post in which palisades closed the open areas between the buildings, but ultimately it became an expansive spread of buildings surrounded by a picket fence.

Cellar holes and rock walls marked most building sites and a stone marker was in the center of the parade ground until 1961 when the marker was moved to the Oneida courthouse. It was feared that the Oahe reservoir would inundate the site, something finally achieved in 1968 when the reservoir water level spread over the Sully table land.

The former hospital is now a granary at a private farm. To reach it from Pierre take US 17 north to the asphalt road at the edge of town. Follow this road to Snake butte, 3 miles. North of the butte bear right onto a gravel road which will pass the hospital building on the left side in 14 miles, in the Okobojo Creek Valley.

Continuing north on this road, turn left in 3

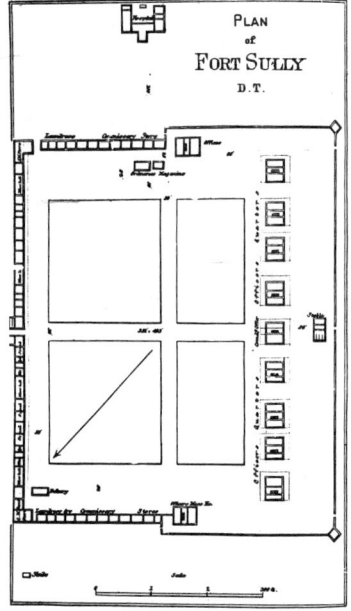

miles at the junction with another gravel road. When the road turns sharply north (right) in 5 miles, stop. From here to the site—or the bluffs overlooking it—is 6 miles along vague traces of a dirt road that can be maneuvered over open range land by a jeep.

Tabeau's Post. 1795-1804. Pierre Antoine Tabeau operated this trading house in or near the Arikara village where he was visited by Lewis and Clark in 1804. Not long after, he left the post for employment with Canadian fur companies. The approximate site is inundated near the east bank of the Missouri about 10 miles north of Mobridge.

Fort Tecumseh. 1822-32. The leading fur post of the Columbia Fur Company in its day, this stockade had the misfortune of being so close

DACOTAH TERRITORY—FORT THOMPSON, ON THE UPPER MISSOURI RIVER, BUILT FOR THE SIOUX AND WINNEBAGO AGENCY.—[SKETCHED BY JOHN NAIRN.]

to the Missouri river that it was flooded out annually. It was turned over to the American Fur Company in 1827 and ceased operation five years later after Fort Pierre had assumed all of its functions and taken most of its worthwhile buildings. From Pierre, go west on US 83 toward the city of Fort Pierre. After turning south on US 83 where it makes the junction with US 14, stop at a point 300 yards south of the junction. This is the site of Fort Tecumseh, bisected by US 83.

Fort Teton I. *Fort LaFramboise I.*

Fort Teton II. 1828-30. *Teton Post.* P.D. Papin and Company built this trading post but it was not long before the American Fur Company had taken it over and moved the property to Fort Tecumseh. The site was near the mouth of the Bad river, generally south of the city of Pierre.

Fort Thompson. 1863-71. *Post at Crow Creek Agency.* A stockade 300 by 400 feet surrounded two dozen buildings manned by two companies of Iowa Volunteers, Fort Thompson was the first military habitation north of Fort Randall during the Civil War. It was turned over to the Indian Agency in 1871. When the stockade was torn down in 1878, its dimensions had expanded to 450 by 650 feet. From Chamberlain go north on South Dakota 47 to a gravel road in about 12 miles that forks to the left (northwest) along the Missouri. About 5 miles along this road is a picnic area which is the easternmost edge of Fort Thompson's area but the post itself has been taken away by the Missouri. Some plumbing pipes from the post hang grotesquely over the water from the eroded banks. This site is about 2 miles south of the new town of Fort Thompson at the intersection of South Dakota 34 and 47W where a marker gives the area's history.

Trudeau's House. 1796-97. *Pawnee House.* This was a trading post operated by school teacher Jean Baptiste Trudeau and 10 men. It was the first house built in South Dakota. The site is uncertain, authorities placing it between four miles north of Pickstown to 15 miles south, due to the erosion and flooding of the Missouri.

Fort Vermillion. 1833-1851. The American Fur Company established this post somewhere to the east of the modern city of Vermillion. Flooding of the Missouri and the passage of time has eliminated all traces but there is a state historical marker on the county road between Elk Point and Vermillion at Burbank. This marker places the trading post to the south on the river bank.

Post Vermillion. 1822-50. *Dickson's Post.* Originally a Columbia Fur Company fort, Vermillion was absorbed by the American Fur Company. It was a small stockade and should not be confused with Fort Vermillion to the east of Vermillion. A marker on South Dakota 50 (south side) to the east of Gayville, 18 miles east of Yanton, tells the story of the post and points to the approximate location, 6 miles south on the Missouri on so-called Audubon Point.

Fort Wadsworth. *Fort Sisseton.*

Post at Whetstone Agency. 1870-72. This was a cottonwood picket stockade with buildings against the inside walls of the stockade and blockhouses at two opposite corners. An estimated 4,500 Indians were at the agency during the active period of the Army post. After the troopers left, the post became a supply depot for the other agencies. Not only has the site been frequently flooded by the Missouri, it is now permanently underwater. The site is about 18 miles northwest of the Fort Randall Dam, in midstream opposite Whetstone creek.

Yankton Stockade. 1862. The Sioux Uprising of 1862 caused the citizens of Yankton to build this 450-foot square stockade at 3d and Broadway. Residents of both Sioux Falls—which became a ghost town at the time—and Yankton gathered at the stockade until the threat was over. The center of the stockade would be modern 3d street and Broadway. At the northeast corner of this junction is the hotel which was the headquarters of General Custer when he was stranded at Yankton by a blizzard in 1873. The city was the location of an Army supply forwarding detachment during the days of the Missouri river posts.

NORTH DAKOTA

FORT TOTTEN *(5)*

Once the life blood of the river trade and the trading and military forts on its banks, the Missouri River has become the executioner of many because of reclamation projects. For this reason it is easy to give directions to the approximate sites; the exact sites no longer matter because they are under many feet of reservoir water.

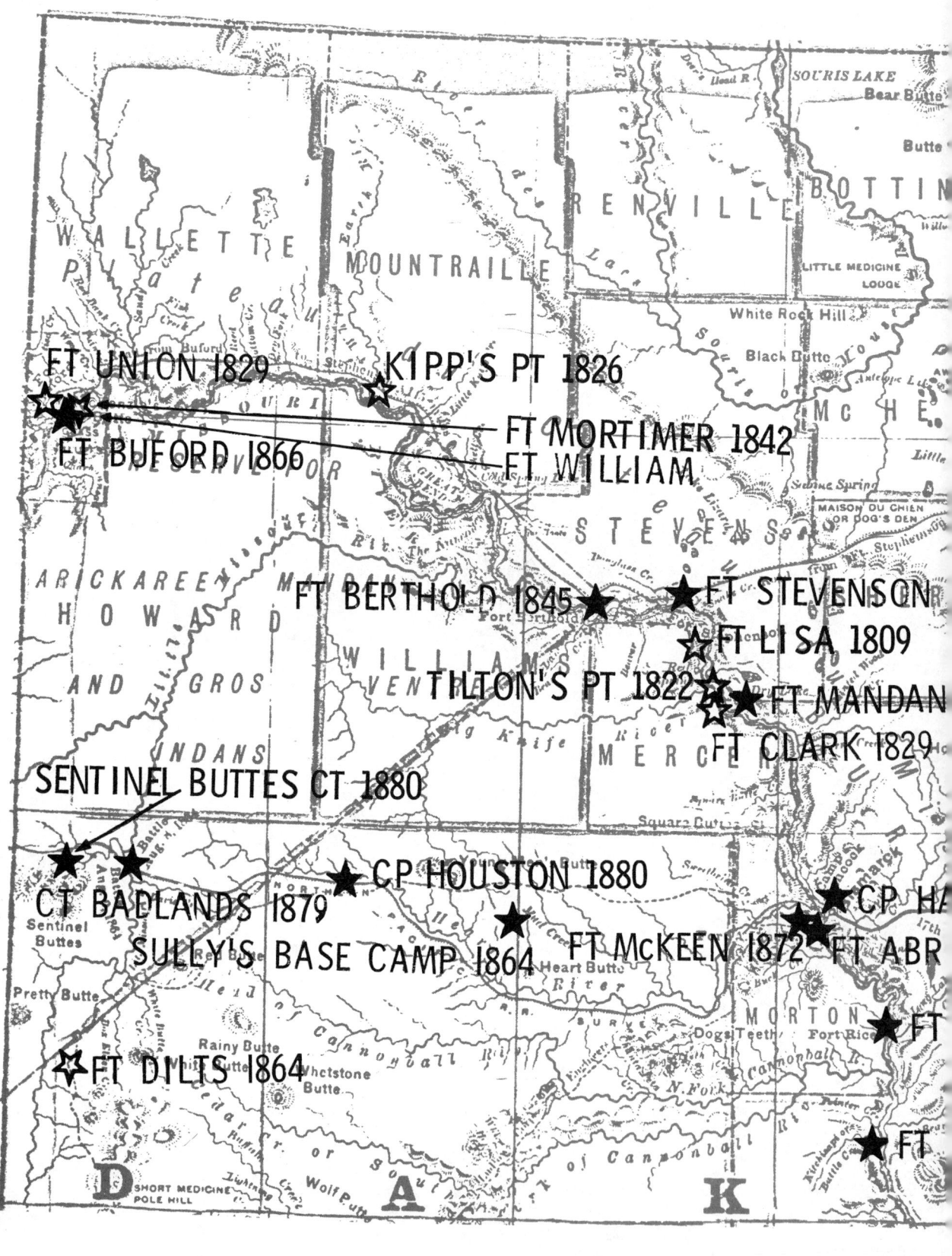

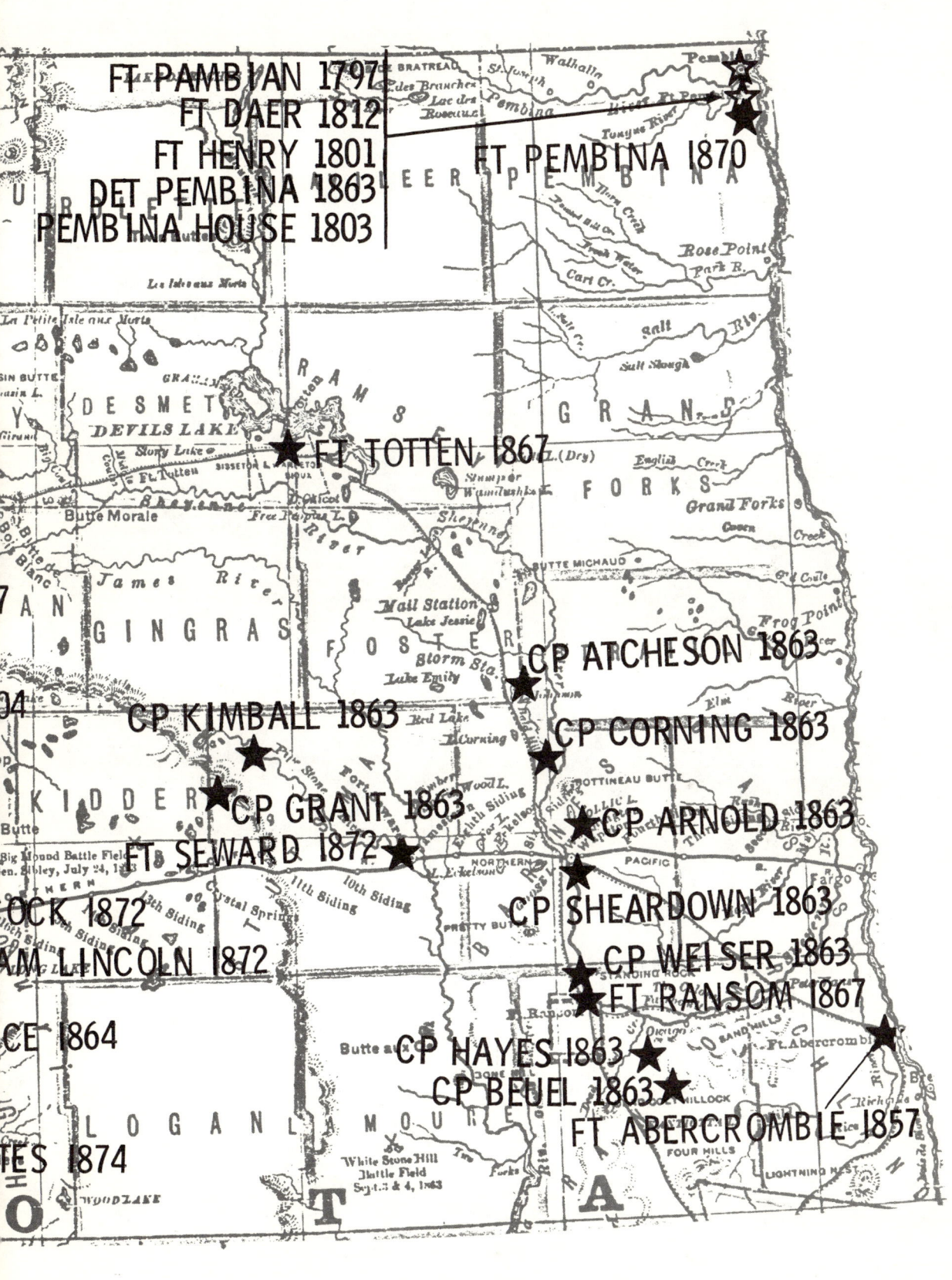

Fort Abercrombie. 1857-77. This post's most critical period was an 1861 siege when an estimated 300 Sioux attacked the post. The fort was saved in a defense that called upon all occupants, military and civilian alike. The post is now a state park with restored blockhouses, guardhouse and stockade at the eastern end of the main street of Abercrombie, a town in southeastern North Dakota, 45 miles south of Fargo on US 81.

Fort Abraham Lincoln. 1872-91. One of the most significant forts of the west, this was the post from which Custer left in 1876 for his ill-fated Battle of the Little Big Horn. Locations of buildings are outlined by stones and identified by markers in what is now a state park and near which a Custer pageant, "Trails West," is presented nightly during the summer season. From Bismarck, take Main avenue (US 10) across Missouri river to Mandan, 4 miles. Turn south at Sixth avenue, go 5 miles to Fort Abraham Lincoln State Park. Museum is on left (east) side of road, main fort site on right.

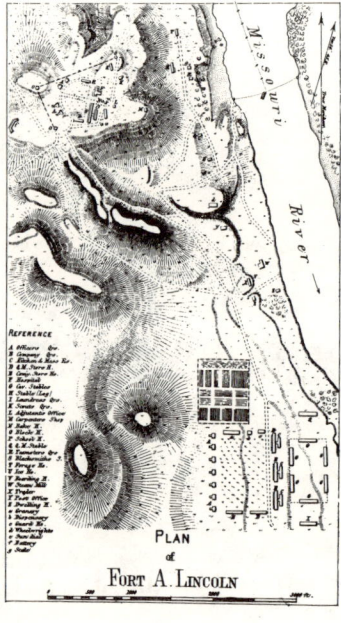

Camp Arnold. 1863. This was a field camp on the return march of General H.H. Sibley's 1863 expedition. It was one of many overnight or short-term camps. Some of the sites have been acquired by the state; only the campsites so identified as Historic Sites will be listed in this *Guide.* Take I-94 west for 49 miles from Fargo to the Oriska exit at North Dakota 32. Go north on 32 for 5 miles to the site, left (west) side of road.

Camp Atcheson. 1863. Sibley's troops were at this camp longer than many because two infantry companies were left here with slower elements of the expedition while the main column rushed ahead after the fleeing Sioux. Trenches and breastworks were prepared for the month-long encampment on the northeastern shore of Lake Sibley; the site of the grave of a private who died here is on a hill to the northeast. Take I-94 east from Jamestown 24 miles to North Dakota 1 exit. Go north on 1 for 53.5 miles to a left turn to Lake Sibley, ½ mile; the campsite was on the northeastern shore.

Fort Atkinson. *Fort Berthold.*

Cantonment Badlands. 1879-83. *Cantonment at Little Missouri Crossing.* Troops were stationed here on the west bank of the Little Missouri river to protect workmen of the Northern Pacific Railroad. From Dickinson, go west on I-94 to Medora exit, 34 miles. Specific directions to the cantonment's site should be requested from National Park Service headquarters 1 mile south of the exit.

Camp Banks. 1863. Sibley's Expedition camped here at what is now called Chaska Historic Site after Indian scout who died during expedition and is buried here. The state-owned park, with marker, is north of Driscoll.

Fort Berthold. 1845-66. *Fort Atkinson.* Fort Berthold was originally a trading post that was joined in 1859 by a fur trading competitor, Fort Atkinson. The American Fur Company, owners of Berthold, bought out the Atkinson venture and moved their operation to the newer post, designating it New Fort Berthold.

The Army arrived in 1864 and, in 1866, moved to the site of Fort Stevenson 17 miles away and for a short time called it "New Fort Berthold," too. All of the sites have been inundated by the Garrison Dam reclamation project and are under Lake Sakakawea.

An overlook of the sites can be reached by going from Bismarck on I-94 for 48 miles to North Dakota 49; head north about 46 miles on this road until it deadends at Lake Sakakawea.

Camp Beuel. 1863. This was one of the first campsites for Sibley's troops after they entered the present boundaries of North Dakota. From Fargo, go south on I-29 about 50 miles to North Dakota 13. Turn westward on 13 about 25 miles to a turnoff to Milnor, right (north). Camp Beuel site is at Milnor.

Fort Buford. 1866-85. This post was one of the most be-deviled forts in the west. It was even the victim of a country-wide rumor that the garrison had been massacred, but there was more imagination than fact to this. The state now manages the site and the remaining buildings. From Williston, go west on US 2 for 7 miles to the town of Buford. The Fort Buford Historic Site is 1 mile south of the town.

Chaboillez Post. *Fort Pambian.*

Fort Clark. 1829-37. Trader James Kipp built this 132-by 147-foot stockade for the American Fur Company. Until closed by a smallpox epidemic in 1837, Clark was one of the most important fur posts of the Missouri. It is now a State Historic Park; no traces of the stockade remain.

FORT BERTHOLD, outside *(1)*

FORT BERTHOLD, inside *(1)*

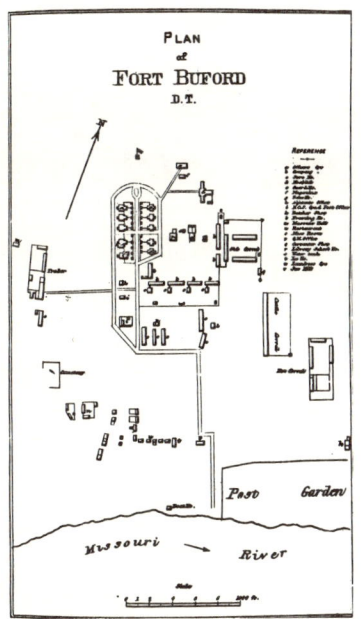

From Bismarck go west on I-94 for 31 miles to New Salem exit at North Dakota 31. Go north on North Dakota 31 for 31 miles to North Dakota 7; turn right (east) for 3 miles, then right again on gravel road (directly east across the river is the site of Fort Mandan, Lewis and Clark wintering-over post).

Fort Clark park is on left (east) side of the road about 6 miles south of the intersection.

Camp Corning. 1863. Sibley's expedition overnighted here on July 16, 1863. From Jamestown, take I-94 east 24 miles to North Dakota 1. Go north on North Dakota 1 for 19 miles to a gravel road south of Dazey. Turn right (east) for 6 miles to an intersection with another gravel road; turn left (north) for about 2 miles to Camp Corning Historic Site.

Fort Cross. *Fort Seward.*

Fort Daer. 1812. This was protection for a group of Scottish and Irish settlers, the first permanent settlement at the mouth of the Pembina river. The site was the same as for Fort Pambian.

Fort Dilts. 1864. This was a sod breastwork in which an immigrant train withstood Indian attacks for 14 days. The lonely site is marked and grave stones memorialize the soldiers killed in the defense. From Bowman, in southwestern North Dakota, go west on US 12 for 18 miles to a hard right turn (north) onto a gravel road. After 2.5 miles, turn left (west) onto single track dirt road. Site is on left (south) side of road, 2 miles further.

Fort George H. Thomas. *Fort Pembina.*

Camp Grant. 1863. Another one of the Sibley expedition campsites, now a State Historic Site. From Jamestown, go northwest on US 52-281 for 35 miles to gravel road (2 miles north of Melville). Turn left (west) for 8 miles—where another Sibley site, Camp Kimball, is north of the road—and then another 8 to a road intersection; Camp Grant site is southwest of this intersection.

Camp Greeley. *Camp Hancock.*

Camp Hancock. 1872-77. *Camp Greeley.* Under the Greeley name this post was intended to protect the railroad construction crews but its mission expanded to supply depot status when the railroad arrived at Bismarck. A single log barracks, 100 by 20 feet, and six other buildings made up this supply point for the Dakota posts, especially Custer's Fort Abraham Lincoln to the south. The site includes a remodeled building, now used as a museum to Dakota pioneers and Indian tribes, at 117 Main avenue in downtown Bismarck.

Camp Hayes. 1863. This was a week-long camp for Sibley's column, giving them time to dig trenches and breastworks—some still in slight evidence—while awaiting supplies and mail. From Valley City, go south on North Dakota 1 for 37 miles south to North Dakota 27. Turn left (east) for 23 miles to a gravel road. Turn right (south) for 2 miles. Turn left (east) for 3 miles to another junction; turn right (south)

FORT MORTIMER
From a drawing by Alexander H. Murray (1)

4.8 miles to the site of Camp Hayes on the first level above the Cheyenne river flood plain.

Heart River Corral. *Sully's Base Camp.*

Fort Henry. 1801-09. Alexander Henry maintained this North West Company trading post at Pembina, immediately north of the mouth of the Pembina river. The site is across Rolette street from the site of Pembina House.

Camp Houston. 1880. This was a temporary camp for the protection of railroad construction crews at Dickinson, south of I-94 on North Dakota 22 in southwestern North Dakota.

Camp Kimball. 1863. Sibley's troops camped here July 22 and 23, 1863, before moving southwest for the Battle of Big Mound. Now a State Historic Site. Directions to the site are at the Camp Grant entry.

Kipp's Post. 1826-30. Fur trader James Kipp built this 96-foot square stockade in order to capitalize on the Indian business until the establishment of Fort Union.

Only slight surface traces indicated the site until the State Historical Society conducted archeological excavations in 1954. These showed that the buildings had been destroyed by fire and, from metal fragments unearthed, had at least one 1-pounder cannon for protection. The latter does not seem to have been much protection: the remnants indicated the cannon had exploded.

Smithsonian Institution River Basin Survey Paper Number 20, publihsed in 1958, gives detailed data on the excavation project. From New Town, go north along Spanish Bay and cross the Little Knife river. Continue bearing left on gravel roads for a total of about 40 miles to a point 1 mile southeast of the mouth of the White Earth river. To the left and probably inundated by the Garrison Dam project is the site of Kipp's Post.

Fort Lewis. *Fort Lisa.*

Fort Lisa. 1809-12; 1823. *Fort Manuel Lisa, Fort Lewis, Fort Vanderburgh.* The Missouri Fur Company built this trading post under the direction of Manuel Lisa and Reuben Lewis, brother of Meriwether Lewis of Lewis and Clark fame, managed the post until it was abandoned in 1812. Under the Vanderburgh name, the post was occupied for a short time by Joshua Pilcher in 1822 or 1823. The site has been inundated by the Garrsion Dam project but the general area can be seen from Pick City near the western edge of the dam.

Cantonment at Little Missouri Crossing. *Cantonment Badlands.*

Fort Mandan. 1804-05. Lewis and Clark built this triangular shaped stockade for the winter of 1804-05; houses made up two sides, pickets the third. It was here that Sakakawea, the Indian girl, joined this expedition. A year later, on the way back from the Pacific coast, the duo noted that all but one house and some pickets had been burned. From Bismarck, take US 85 north for 40 miles to Washburn. From Washburn take a gravel road to the west 14 miles to Fort Mandan State Park, the approximate site of the post, where a replica has been built by the McLean County Historical Society.

Fort Manuel Lisa. *Fort Lisa.*

Fort McKeen. 1872-91. On a bluff 270 feet above Fort Abraham Lincoln, this stockaded post preceded the larger camp and, after completion of Abraham Lincoln, became a sub-post of it. There are reconstructed blockhouses and marked building sites at the location, now part of the Fort Abraham Lincoln State Park. Directions to the site are in the entry for Fort Abraham Lincoln.

Fort Mortimer. 1842-46. Fox, Livingston and Company built this trading post near the remnants of Fort William as competition for the American Fur Campany's Fort Union, three miles to the west. In three years it sold out to the competition. An adobe trading post was on the same spot in 1858; the Army built Fort Buford here in 1866. Follow directions to Fort Buford; the sites are almost the same.

Fort Pambian. 1797-98. *Chaboillez Post.* The first trading post in North Dakota, this was a rough camp built by Charles Chaboillez of the North West Company. Four canoes full of furs were the season's total of trade from the Indians, usually received in barter for rum. The buildings were burned in 1815. The site is on the south side of the mouth of the Pembina river, now Selkirk Park on Stutsman street in Pembina (on I-29 about 3 miles south of the Canadian border).

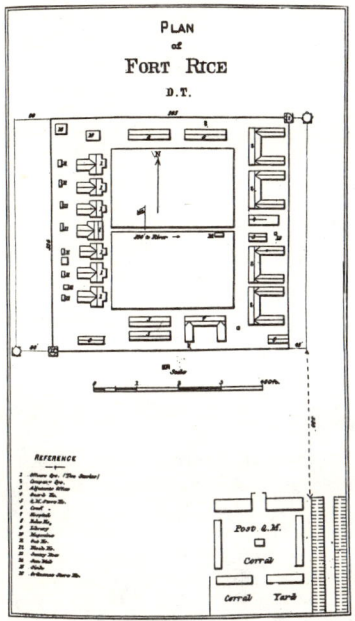

Although the stockade and all buildings have disappeared, mounds and impressions on the surface indicate the layout of the post.

Take I-94 west for 48 miles to North Dakota 32. Turn left, going south on 32 for 20 miles to North Dakota 46. Turn right (west) for 2 miles to a gravel road on the left. Turn left (south) and follow this curving road for about 15 miles to the village of Fort Ransom. The fort site is in a valley south of the Sheyenne river on the left side of the road 1 mile south of the town.

Fort Rice. 1864-78. Most of the major expeditions between 1864 to 1873 started from Fort Rice, a stockaded post that commanded most of the Missouri river in the Dakotas during its early days. Several blockhouses have been reconstructed and building sites are marked in a state park, though one burned in 1977. From Bismarck, proceed to Fort Abraham Lincoln and then continue south for 30 miles to the town of Fort Rice. Turn left (west), cross railroad tracks, bear right one mile to the fort site overlooking the Missouri river.

Cantonment at Sentinel Butte Station. 1880-82. Troops protected the Northern Pacific construction crews in the shadows of the strategic Sentinel Buttes beginning in late 1880 and periodically for a couple of years. From Cantonment Badlands continue west on I-94 to Sentinel Butte exit; turn south to the railroad, approximate site of the cantonment.

Fort Pembina. 1870-95. *Fort George H. Thomas.* Last of the several fort-type places in the Pembina area, this Army post had facilities for more than 1,000 soldiers. It was auctioned off in 1902, providing the area with several evidences of Army presence: a renovated officers quarters in the town of Pembina and water hydrants and mains in the town water system. The site of the post cemetery still is used. From Pembina, go south on I-29 for 3 miles to fort site adjacent to Pembina airport.

Fort Pembina. *Pembina House.*

Pembina House. 1803-23. *Fort Pembina.* This Hudson Bay Post operated on the north bank of the mouth of the Pembina river until it was determined that the site was in the United States rather than Canada. The site is on Rolette street where Minnesota Volunteers were stationed in 1863 (Pembina Detachment), in Pembina on I-29 south of the Canadian border.

Detachment at Pembina. 1863-64. Minnesota Volunteers set up a temporary cantonment at the site of Pembina House to protect the settlers during the so-called Sioux Uprising. A marker on the site records the fact that members of the garrison organized North Dakota's first Masonic lodge during their duty here.

Fort Ransom. 1867-72. Protecting the railroad was a major concern at this post and frequently only a skeleton guard was left behind to protect the families and government property. This stockaded fort is now a state park.

FORT RICE, 1865 *(6)*

FORT YATES (1)

Fort Seward. 1872-77. *Fort Cross, Fort Sykes.* A detachment of infantrymen arrived at this site in 1871, built a small building for shelter, and began to guard the railroad construction crews. The permanent post included the area of this shelter and, 100 yards north and atop a bluff, the rectangular main post, including a two-company barracks 230-feet long. The modern site is marked by a cannon on the bluff northwest of Jamestown about 2 miles on US 52-281.

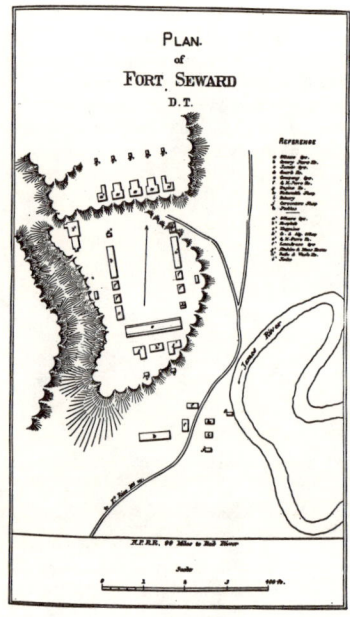

Camp Sheardown. 1863. Rifle pits at this Historic Site are the only traces of Sibley expedition use on their way to engage the Sioux. From Valley City, on I-94, go south about 3 miles on a gravel road to the marked site, left (east) side of road.

Fort at Standing Rock Agency. *Fort Yates.*

Fort Stevenson. 1867-83. *New Fort Berthold.* Although given the latter name when it was first occupied by troopers from Fort Berthold, the Stevenson name soon was adopted. It was an important post because of its strategic location, both along the Missouri river and on the trans-Dakota east-west trail.

The post was quadrangular in shape, buildings or fences surrounding a parade ground that was flanked by 10 buildings. The post was inundated by the Garrison Dam project.

To reach an overlook of the general area, take US 83 from Bismarck for 73 miles north to North Dakota 37. Turn left (west) on 37 for 6 miles to Garrison. Three miles south of Garrsion is the Fort Stevenson Recreation Area, overlooking the approximate site of Fort Stevenson.

Sully Base Camp. 1864. *Heart River Corral.* The forwardmost base of Sully's 4,000-man expedition in 1864 was this camp on the Heart River, a 10-day post manned by 125 soldiers, 250 emigrants and a decoy cannon. A stone marker in the center of a pasture memorializes the Army use for the Battle of Killdeer Mountain. From Dickinson, go east on I-94 for 27 miles to Antelope exit. Go south on gravel road for 14.3 miles. Marker can be seen in field to the left (east) of the road just before crossing the Heart river bridge.

Fort Sykes. *Fort Seward.*

Tilton's Post. 1822-23. This was another post built by James Kipp. It was just about the Fort Clark site and was soon abandoned because of

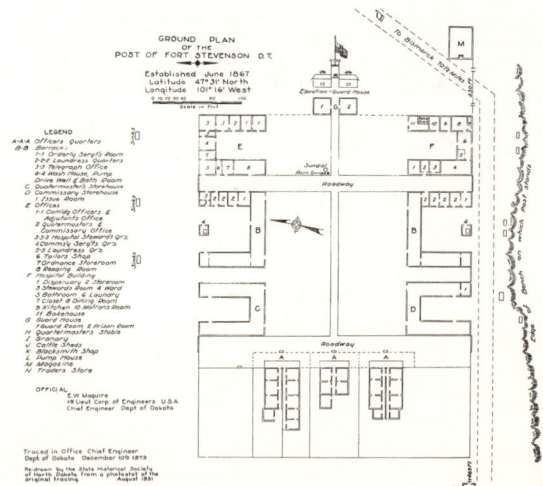

Indian hostility — one of the employees was killed at the fort's entrance, in fact. The approximate site can be reached by following the directions to Fort Clark; Tilton's Post was a mile or so north of Fort Clark but there is no trace. Kipp contributed to the deterioration of the site by tearing down many of the pickets and using them in a trading post he operated across the river at the Mandan Villages during the winter of 1825-26.

Fort Totten. 1867-90. One of the best preserved of the forts of the old west, Fort Totten's period photographs are hard to distinguish from modern photography. Almost all of the Army's brick buildings are left around the parade ground; the only difference is that the saplings planted 100 years ago are now substantial trees. The state has made a park of the parade ground and its flanking 15-plus buildings. From Devil's Lake, take North Dakota 20 south for 5 miles to North Dakota 57. Turn right (west) on 57, following it around to Fort Totten Indian Agency and State Park. The town of Fort Totten is immediately north of the park and fort site.

Fort Union. 1829-1865. Kenneth McKenzie built this 220- by 240-foot stockaded trading post to be the model fur post of the Missouri. It overcame all opposition during its heyday until the Army clamped down on illicit liquor bartering. Troops were stationed at the fort in 1864 until it was closed in mid-1865. The National Park Service is developing the site at the Fort Union Trading Post Historic Site. Follow directions to Fort Buford; take the gravel road west from the town west to the Montana border. Fort Union is on the Dakota side, 50 yards south of the road.

Fort Vanderburgh. *Fort Lisa.*

Camp Weiser. 1863. The Sibley column overnighted here July 13-14, 1863, naming the site in honor of the surgeon of the First Minnesota Mounted Rangers who was later killed at the Battle of Big Mound. From Valley City, go south on North Dakota 1 for 24 miles to North Dakota 46. Turn left (east) for 9 miles to a dirt road. Turn left (north) on dirt road 1 mile to the Historic Site.

Fort Yates. 1874-1903. *Post at Standing Rock Agency.* Although this post was established in order to watch over the Indian agency, it was here that the Messiah Craze and ultimately the Messiah War began. Some Army buildings and the cemetery remain, mixed with buildings of the agency; the Standing Rock is mounted on a bluff at the north end of the town. From Mobridge, S.D., take US 12 cross Missouri river for 34 miles northwest to McLaughlin. Turn right on South Dakota 63 for 9 miles to the state line where the route becomes North Dakota 6. Six miles north of the line, turn right (east) on North Dakota 24. Follow this gravel road to its end at the Standing Rock Agency and town of Fort Yates.

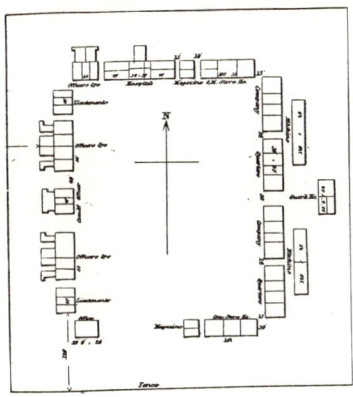

PLAN of FORT TOTTEN D.T.

FORT UNION *(1)*